Beyond Chaos

Also by Mark Ward

Virtual Organisms:
The Startling World of Artificial Life

Mark Ward

Beyond Chaos

The Underlying Theory Behind Life,
the Universe, and Everything

Thomas Dunne Books
St. Martin's Press ⋈ New York

THOMAS DUNNE BOOKS.
An imprint of St. Martin's Press.

www.stmartins.com

ISBN 0-312-27489-0

First published in Great Britain by Macmillan
under the title *Universality*

First U.S. Edition: July 2002

10 9 8 7 6 5 4 3 2 1

For Clare, forever

Contents

Preface

Introduction 1

Chapter One
The Wonder of U
9

Chapter Two
The Littoral Truth
43

Chapter Three
The Midwife of Creation
88

Chapter Four
The Rhythm of Life
138

Chapter Five
The Business of Complexity
177

Chapter Six
Me and U
219

Contents

Chapter Seven
Just U
274

Notes **299**

Bibliography **309**

Index **317**

Preface

This is a little book about some very big problems. Some of the biggest ones there are, in fact. The problems that give our existence its depth and drama, its details and delights. The problems that, in short, bring our lives to life.

Physics, with its concentration on inert lumps of matter, typically has little to say about these things; it is only people who shout about stock market slumps, complain about the buses being late, or the way their heart flutters when they see a pretty face. But of late physics is turning its attention to these matters and finding that it can say quite a lot about them, after all.

This book attempts to trace the history of this new field of research, which goes by the grand name of Universality, and show just how those inert lumps of matter conspire and self-organize to produce life with all its associated frustrations, desires and disasters.

However, one of the trickiest aspects of writing about this field is the potentially vast scope of the subject. Although Universality does not encompass everything, sometimes it seems to come close. This made the hardest job of writing any book, deciding what to leave out, all the more difficult.

Inside are the key experiments, theories and thinkers working on this subject, and this book reflects the state of play in the field. I've tried to include everything that is important and relevant without swamping the text with references and asides. Hopefully this picking and choosing will mean anyone reading this takes away a willingness to look, stop and see the forces at work in your life and the wider world. Forces that are often beyond sight but not understanding. It is perhaps too grand to say that it is a vade mecum to a new way of seeing the world, but it might make you stop and wonder occasionally. Welcome to your life once again.

In an ideal world a book about self-organizing systems would write itself. All I would have to do is put all the references, papers and reports together inside a bag, shake them up and then tip out the finished text. Unfortunately, it took a bit more work than that and I have to thank lots of scientists and researchers who took the time to answer my incessant questions, and always responded to my nagging with courtesy and kindness. Thanks all who helped, especially David Avnir, Alberto Laslo Barabasi, Marcia Barbosa, Alastair Bruce, Winn Farrell, Ary Goldberger, Mel Goodale, Henrik Jeldtoft Jensen, Leo Kadanoff, Scott Kelso, Mark Newman, Sverrir Olaffson, Tom Ray, Yonathan Shapir, Matthew Smith, Ricard Solé, Mark Venables, and Heiko Weissbach. I should extend special thanks to the indefatigable Gene Stanley and Per Bak for spending lots of time explaining in simple terms the intricacies of Universality.

The authors and publishers of quotations and extracts are acknowledged in the notes and references at the back of the book.

Extra thanks are also due to Matt and Monica Findall-Hawkins for letting me and Clare share their elegant beach-front apartment in Spain. I couldn't have done Chapter Two without you. It was a scorcher. Also double plus thanks to Matt for the tip about After Such Knowledge.

I also need to thank South West Trains which consistently delayed the train services into Waterloo and unwittingly helped me keep up with my reading.

Chris Stewart and Julian Alexander deserve lots of credit for seeing the potential in my rough ideas and negotiating the deal with Macmillan. Catherine Whitaker at Macmillan should get a medal for putting up with me and persevering in her attempts to show me how to turn my jumbled writings into something much more readable.

Finally, I have to thank Clare for her unfailing patience and generosity while I wrote this book. Her clear thinking and concern constantly show what a treasure she is and how lucky I am.

Mark Ward
Surrey

Introduction

People fear nothingness, and dread its approach. But unlike other threats, this one can never be stayed, nor fled, not in the least, not even if one were given all the powers of heaven. Nothing gives it an instant's pause; nothing can.

Metaphysics, Richard Taylor

The High Priests of telescopes and cyclotrons
keep making pronouncements about happenings
on scales too gigantic or dwarfish
to be noticed by our native senses

'Ode to Terminus', W. H. Auden

There is an invisible force at work in the world that is nameless and ubiquitous. You cannot reach out and touch it, but every day you will feel its influence. You may often see hints of it at work, or suspect its presence, but you have never been able to pin it down. It works beyond sight and hearing yet you encounter it on a daily basis. No matter how often it touches your life you probably have no real inkling of its reach, nor how it connects you to the larger universe. It has a dynamic that is impossible to predict, control or stop, and it has much of the universe in its thrall. Its ascendancy over the earth was assured before the planet had even formed.

If the influence of this hidden power were revealed to you for just a moment, you would see its tendrils everywhere. It has given the world its structure and ordered the organisms upon it. It is at once pitilessly powerful and almost unbearably gentle. It has scattered the stars across the sky and made some of them glimmer; it routinely shapes mountains and levels cities. Yet it can also make mice, form flowers and even teach you how to walk.

Every day you feel its influence at second hand as it dictates the clothes you want to wear, the value of the money in your pocket and whether you will get to the office on time. Your body and the world beyond it beats to its rhythm. However, before now – before you opened this book – the truth has been hidden. You were right to be suspicious, because important information has been kept from you and almost everyone you know. This information has not been denied you because of a conspiracy – a government hiding the facts or aliens obscuring the evidence. Not even a grand alliance between the two is responsible. Reality is more mundane than that. You've not been told because before this age of marvels no one knew. But they do now.

Within these pages *Universality* will be laid bare and its reach defined. Its dynamics will be dissected and laid out for your inspection. You will have the scales lifted from your eyes and confront the mystery mankind has sought to solve since the day he first could think.

This is a mystery that physicists have long concerned themselves with, but only in the last few years have there been significant breakthroughs into the nature and extent of Universality. The physicists doing this remarkable work are not the ones working on 'Theories of Everything'. Those physicists are probing matter ever more deeply to see how everything fits together: they are searching for the synthesis of everything that exists. The physicists working on Universality are concerned with problems much more important than that.

The search for a Theory of Everything has become one of the most urgent pursuits of our times. Fuelled by a steady trickle of research, lots of books and many more newspaper articles, it has become an involving, international project that has swept everyone up in its drama and promise. Given the momentous progress that physics and physicists have made over the last century, not to mention the money spent on this project, you could be forgiven for thinking that the search is nearly over and that the grand unity is at hand.

Until recently there has been precious little evidence that the greater knowledge of this large and small spilling from the wake of these grand physics project was putting us any closer to answers or explanations of how complex structures arise, not only complex physical structures such as the splash of stars in the night sky and rugged mountains, but also complex social structures such as the roll of history and the boom and bust of the stock markets. Physicists studying the elements of matter have little to say about these matters. In contrast, physicists studying Universality are consumed with them. Life, from its beginnings to its end, is shot through with Universality. We, and everything we take part in or react to, are no more than echoes of that call. These are the secrets that Universality can explain.

Every year brings a greater understanding of all the parts to be found in living and dead matter, but the question of how they interact to bring forth you and me, and a lot of other things besides, remains unanswered.

Particle smashers are getting bigger all the time and the collisions carried out in their dark hearts are revealing that fundamental particles are not quite as fundamental as we thought. Organs, cells, genes and proteins are being probed ever more closely and all the ways they can combine and react are being logged. Now it is becoming obvious that the answers to these questions are never going to be found where the particle physicists are looking. This is because another group of physicists has found the answers elsewhere already. And we keep asking the question: why is my life like this?

Telescopes and microscopes are extending the limits of our vision and we know more than we ever did about our physical make-up. But the detail that the physicists have discovered has given precious little insight into the dynamics of the blooming, scruffy complexity of the world. Stubbornly the central mystery remains: what gives rise to this order and structure?

In many ways the world is coming to resemble a vast work of art, something fantastic and frantic like Hieronymous Bosch's *Garden of Earthly Delights* perhaps. The closer you get to the canvas, the more detail is revealed and the greater is your appreciation of the skill it took to apply the paint.

This scrutiny is giving us more than just a better appreciation of the diversity of the world. It is starting to turn up some curious regularities. Similarities are being found in the pulsing of quasars, global weather movements, and the branching networks of nerves and blood vessels in bodies. You can see it at work in the head of a cauliflower and the face of a pretty girl. You might see it every day in the bathroom mirror or pulling clouds into characteristic shapes. You can even hear it in the music of Bach and the beating of your heart. It shows that snowflakes have memories but fashion is forgetful. These are vastly different systems, yet they seem to share some common principle, a single dynamic, a universal affinity.

Physics has a problem with complex structures such as pretty faces and populations. Physics, and most of the rest of science, has little to say about how most of the world got to be like it is. Physics may be tackling what is too small and what is too large, but it has shied away from the mundane and everyday. This is not because the shapes of clouds are easy to explain but because they are very difficult to explain. It has taken a long time for physics to work out what was going on in a cloud and what was driving those characteristic shapes. Now a lot of people have a really good idea about what is going on, and physics is coming home.

We live our lives between the tiny and the titanic. Smaller than galaxies but larger than cells, we inhabit the world that physicists have, until now, struggled to say anything meaningful about. All of a sudden the dynamics of order and complexity are being laid bare. Questions that have remained unanswered for centuries are receiving a reply. And it is one that makes sense and might bring some much needed relief to anyone who cares to listen or read about it. We need no longer be comfortless: our pain has a name.

Thankfully it turns out that we aren't doomed to eternal ignorance and bafflement and, what is even better, insights into the workings of the universe that matter to you and me are not closed off to all but a high-priesthood of particle physicists. You don't have to spend years as a practising physicist to gain a profound insight into much of the universe's workings and those events that frustrate us on a daily basis. Just by living your life you have as much experience with Universality as the physicists who study it for a living.

This is why the ideas in this book are more important than a Theory of Everything. They matter more because they help explain many of the events that baffle and frustrate you and me on a daily basis. They also reveal why much of the world looks like it does.

They might even help answer the oldest questions of all. Not questions about particles and quarks, but questions about people and society. The questions all men have asked since they could frame a question: Where do I fit in? How does it all make sense? Does life have any meaning? We have always wanted to understand ourselves better – restless stubborn people that we are – and now we have a much better chance of seeing where we stand, why it makes enough sense, and how to pluck sense from all of this. At the very least they might give you a new perspective on life and a whole new set of questions to mull over.

A consensus is emerging about how complex structures emerge, grow and sustain themselves. Phenomena that were thought to be unique are turning out to have a great deal in common. The theories that explain how complex structures emerge and maintain themselves are supremely relevant. They will be far more useful than a Theory of Everything that offers an answer to a question you can't ask. These theories provide insights into the dynamics of the most pervasive forces and events that shape us and our lives. These ideas will not give you control, but they will help you understand and be at ease with the world. They will show you how you fit in and give you a better chance of understanding, or accepting, how the world works.

With breathtaking simplicity these theories are laying bare life. They are tackling head on just why some events happen irregularly and others never do. Prior to these breakthroughs it was thought that each complex system, be it a pulsar or pulse pattern, would need its own theory to explain it. Now there is one that explains them all. Universality and unity is almost all.

These insights into complexity are not going to give you a simple answer, but in the long run they might confer something more useful than that. They might help you understand the way of the world better and accept some of the things that happen to you, though you might not be happy to hear some of the implications.

These theories imply that life is not random, cruel and uncaring. Rather there is a hidden rhythm to the run of our lives that is also found at every scale throughout the universe. At every level, from the cellular through the cerebral to the continental and cosmic, the same threnody is pulsing.

Take a look around. This stuff is everywhere.

Chapter One

The Wonder of U

Behold now, standing before you, the man who has
pierced the air and penetrated the sky, wended his way
amongst the stars and overpassed the margins of the
world, who has broken down those imaginary divisions
. . . which are described in the false mathematics of blind
and popular philosophy. By the light of sense and reason,
with the key of most diligent enquiry, he has thrown
wide those doors of truth which it is within our power
to open and stripped the veils and coverings from the
face of nature.

The Ash Wednesday Supper, Giordano Bruno

If you have abandoned one faith, do not abandon all
faith. There is always an alternative to the faith we lose.

The Comedians, Graham Greene

An English Magus

It is 28 May 1583. Summer is beginning to take up its
lease. At Mortlake, a village to the west of London, in a

house sitting between the parish church and the river, Dr John Dee is communing with angels.

Candles flicker in the dim, book-lined study where Dr Dee sits waiting for the angels. Edward Kelley, the medium who can draw the angels to the room, is sitting at a table set between Dee and the hearth. Before Kelley, on a cloth adorned with occult symbols, stands a skrying glass of polished obsidian. Through it he sees into the spirit world. The signs upon the cloth have been drawn according to instructions vouchsafed to the pair in previous seances.

As he watches Kelley prepare himself Dee wonders what secrets will be imparted this evening. He is an eager, unquenchably curious man and embarked on this course of action purely to expand his knowledge of the universe, to advance upon its greatest secrets, and help him ascend through the celestial spheres and contemplate the very face of God.

Kelley is peering intently into the glass, waiting for the golden curtain he can see within to be drawn back to reveal the figures and scenes beyond. Kelley has eyes keener than any man Dee has known: no other medium can call spirits to the shew stone in the numbers and varieties that Kelley can.

Though Dr Dee's knowledge of the esoteric and occult dwarfs that of his skryer, he can see nothing in the cloudy glass nor tempt anything to it. Dee may have it from the greatest authorities that the universe is suffused with the servants of the Lord like air bubbles in a mullion window but, though he knows they are there, he cannot see them. Any magus making a journey through the spheres would encounter their many different forms – Decans, Thrones, Virtues, Dominions, Seraphim, and many more. Despite the bustling numbers that crowd the empyrean halls of heaven Dee is as a blind man to them.

Dr Dee does not only deal with angels. *Inter alia* he is an alchemist, a mechanician, a mathematician and an astrologer. And now, last of all, he has become a summoner of angels – a magus. His broad interests and piercing intellect will lead men writing about Dee years later to call him 'an ornament of the age'.[1]

The good Dr Dee has long been acquainted with magic. He studied at Louvain, where years before the infamous alchemist and magus Henry Cornelius Agrippa was employed by Margaret of Austria. Even though the man had been dead for over a decade the place still rang with his influence. Dee even owns a copy of Agrippa's infamous survey of Renaissance magic *De occulta philosophia*.

Now Dee and Kelley regularly spend hours trying to call angels to the shew stone or tease information from those that appear. They spend as long trying to decipher the elliptical and cryptic language used by the angels. The good spirits rarely say anything, plainly preferring to speak in obtuse metaphors and allegories, perhaps for fear of giving away too much to mere mortals.[2]

On this warm May evening the seance starts well. Even before the shimmering curtain within the shew stone is drawn back Kelley says that a spirit has joined them in the chamber. Dr Dee looks up from his papers and peers intently around the room. The shapes in the shadows cast by the candles promise much but reveal nothing to him. He follows Kelley's gaze but can see nothing, not even when Kelley points and tells him where to look.

Discouraged, the Doctor bends again to his writing and gruffly asks Kelley to tell him what the spirit is doing. The 'spiritual creature', says Kelley, has taken the form of a 'pretty girl aged between 7 and 9, wearing a gown of sey [a soft, finely woven woollen fabric] in changeable green and red'. She topples some books in Dee's oratory as she moves around the curtained nook that he uses for his private worship.³

Kelley suddenly looks over his shoulder at a perspective glass propped against some bookshelves in one corner. Dee asks him what is happening. He says other angels are using the glass to speak to the spirit-child and are admonishing it for its clumsy antics. The rebuke reveals to them the name of the child: Madimi. They leave the spirits to argue and break off to take their supper.

When they return to the chamber, Kelley says it is quiet – the spirits have departed for now. He struggles to draw a presence to the stone, and then suddenly sits back so sharply he almost upsets his seat. He says a spirit is standing before him upon the table. He describes its appearance for the Doctor: 'like a Husbandman all in red apparel, red hose close to his legs, red buttoned cap on his head, yea, and red shoes.'⁴

Without prompting, the spirit, which is called Murifri, delivers cryptic prophecies to them: 'The Waters pour forth weepings, and have not moisture sufficient to quench their own sorrows' and 'The Bodies above are ready to say, We are weary of our courses.'[5]

This warm May evening marks a turning point for the spiritual conferences. After this date the spirits appear far more frequently and Kelley can see them in the chamber of practice, not just through the lens of the stone. Over the coming weeks and years a succession of spirits make themselves known. Jubanladace, Gabriel, Raphael, Uriel, Nalvage, Panlacarp, Aphlafben and Levanael[6] become familiar presences to the pair.

Many of them are given to portentous predictions. One called ATH is preceded by the roaring of a great voice around the shew stone. She declares: 'Ignorance was the nakednesse wherewithal you were first tormented, and the first Plague that fell unto man was the want of Science.'[7]

Angels and Insights

Summoning angels by the hearth may seem far removed from the glittering, almost science-fictional physics practised today but, if truth be told, the two have a lot in common. It may be true that Dee's use of a scrying glass, seance and spell book is vastly different to the titanic engines employed by modern physicists, but an affinity does exist between the two pursuits.

Take away the tools that Renaissance magicians and modern physicists employ and you will find they share more than beards and boundless curiosity. The similarities begin with the fact that both are tackling the same problem: how to draw back the veil of nature and reveal the naked face of the world beneath. More important than this is the fact that Dee undertook his angelic conferences with the same attitude that spurs contemporary physicists to do their esoteric work. This common cause persists across the ages and transcends the subjects studied by Dee, twentieth-century physicists, and even their ancient forebears. It binds them all and explains why they began their work in the first place. It renders irrelevant any of the methods chosen to tackle a problem and, when you consider the progress they have made, lumps all the academics, philosophers and scientists together into an ignorant huddle.

To get a better understanding of what it is and why very different professions might answer its call we need to go back to John Dee. We need to find out why he spent the best part of a decade talking to angels and just what he was trying to accomplish.

History has not judged John Dee kindly. In his day Dee was an intellectual force to be reckoned with, a scholar renowned near and far. He was tutor to Robert Dudley, later Earl of Leicester, and to Sir Philip Sidney, the author of *Arcadia* and *Astrophel and Stella*. In later life he acted as unofficial philosophical adviser to the august social circle that surrounded these two great men.

Dee's house served as an unofficial academy and he gladly housed the scholars and students who came to visit him and learn from him. The students were well served with books because he owned the largest library in Elizabethan England. Dee reclaimed many of the texts scattered by the spoliation of the monasteries brought about by Henry VIII, and he owned many rare and ancient writings found in few other places. The journals of men who visited his house say that texts, papers and scrolls were piled everywhere. No other librarian was as ready to share his treasures as the good Dr Dee.

He was an ambassador of learning, a charismatic and enthusiastic scholar who encouraged many lesser men, who in turn went on to spread the fruits of the Renaissance throughout Britain. He wrote a celebrated preface to the first English translation of Euclid and lectured on mathematics in Paris and Louvain. Philosophers from Bohemia and Denmark travelled to England just to talk to him and to find out his opinion on whatever interested them. They rarely went away dissatisfied.

Queen Elizabeth counted him as one of her most trusted advisers. It was Dr Dee who picked 14 January 1559 as the day of her coronation. That the dominion of England and its queen grew almost unchecked afterwards only added to his standing among the wisest men of the commonwealth. He counted the greatest geographers, cartographers and mathematicians in Europe as his friends and correspondents. The Renaissance produced more geniuses than most ages but Dee could stand shoulder to shoulder with them all.

Today the view of him is very different. If people have heard of him at all, they think of him as a dabbler in the occult arts, a wicked man who forged pacts with demons. Worse, they may see him as a gullible old man deluded by his eagerness to believe the fantasies that Edward Kelley spun, fabrications that any reasonable man would see through in an instant.

However, if Dee was deluded then so were most of the scientists and philosophers in Europe. He was not striking out on his own when summoning angels; he was simply applying the knowledge available to any educated man of his time. The books he used could be found in the library of any scholar and the methods he employed had been tried and tested by many before him. At all times he was drawing on a widespread and well-known philosophy. As far as he was aware, he was applying the most ancient and purest knowledge to the greatest of tasks – the salvation of mankind.

Life Lessons

Given the times in which Dee lived, it is no surprise that he, or anyone else, thought mankind needed salvation. The fifteenth and sixteenth centuries were times of enormous turmoil. Every aspect of human life underwent profound change. Feudal social relationships that had persisted for centuries broke down as the ideas of nations, monarchs and subjects replaced those of lords and vassals. The growth of towns and cities changed ideas of how society could, and should, be ordered. An entire way of life with all its regularities and verities suddenly became outdated and irrelevant.

At the same time the certainties of the Catholic religion were being swept away by the Reformation.

Nothing could be the same after 1517 when Martin Luther nailed his ninety-five theses to the door of All Saints' Church in Wittenberg. Everyone heard about what might otherwise have been a little local theological dispute because now, thanks to Gutenberg's movable type, the word could be printed quickly and distributed far and wide. The advent of printing also meant that people could read the Bible and interpret scripture themselves.

Tyndale's translations of the Bible from Greek to English meant that scripture could be read to, or by, anyone that was literate and interested – and increasing numbers were. Many of the protestants had their own copy of the Bible and preferred to interpret what they read in their own way. Their vocal protests about how the Bible should be interpreted earned them the name they bear to this day. As a result heresies, sects and alternative forms of worship sprung up everywhere.

In Britain the shifting affections of Henry VIII and the subsequent spoliation of the monasteries saw the creation of the Church of England almost overnight. The king did what was previously thought unthinkable by openly defying the edict of the pope, God's representative on earth.

In these secular times it is hard to grasp now what these sweeping changes must have meant to people who saw everything in religious terms. Religion gave shape to their world: it defined how they should act and where they fitted in. It must have seemed as if the very earth was shifting on its axis.

The reformation of the spirit was accompanied by a

reformation of the heavens. Copernicus' *On the Revolution of the Heavenly Spheres* was first published in 1543 – shortly after the death of its author. It did away with the accepted Aristotelian vision of the universe and put the sun at the centre of the action. Earth became just one planet among six orbiting the sun. Shockingly this meant that mankind was turned into a spectator rather than the main attraction.

Europe itself seemed to shrink as voyages of discovery were made and fantastic lands discovered. The world expanded, as Europe was shown to be smaller than many of the new lands. Mental horizons had to expand to cope with a suddenly larger globe. For people used to a pace of life measured by the seasons and the plod of the plough-horse these changes must have come at a dizzying rate. Every aspect of life was either changed out of all recognition or called into question.

Possibly because they lived through such turbulent times and so much in their lives went unexplained, the men and women of late Tudor and early Renaissance times were happier when everything was in its place and obeying what they viewed as natural laws. They regarded chaos and disorder with horror and a sign that the universe was unbalanced. The world was physically sensitive to the actions of men and would react accordingly if their conduct was wanting. Laymen and clergy alike collected stories revealing that this was the way the world worked. Puritans were particularly assiduous about collecting examples of what happened to sabbath-breakers.

The popular conception of the universe held that everything in creation had a place ordained by God. This

'Great Chain of Being' defined and ordered the natural world. It arranged everything in a hierarchy and gave it a role and a set of obligations and expectations. Chaos and disorder meant that the cosmic scheme was upset and natural laws were being disobeyed. Nothing good could come of such turmoil. With social, religious and spiritual changes crowding one on another, there was a crying need for order to be restored. It is no wonder that many scholars, such as Dee, were trying to find a way to return to better times.

He wasn't alone. While the larger world was arranged as the Great Chain of Being, the play of historical events was seen as a perpetual movement away from an ancient shining golden age to the dull leaden days of the present. Scholars and clergy saw themselves as assailed on all sides by immorality and licentiousness. They ached for a return to more holy times. The chaos fracturing Renaissance life was a reminder of how far removed contemporary life was from those better, balanced times.

So when scholars looked for a way to fix present problems, they looked back to a time when everything was in its place and mankind was beloved of God. This perpetual desire to return to purer, more harmonious times gave the Renaissance (literally rebirth) its emotional impulse and its reforming zeal.

Mining the Past

It is fortunate then that as scholars were looking for guides to how the ancients used to live the texts that could help them were rediscovered. It is hard to know which came first – the desire for reform or the ancient texts – but the result when they were mixed was explosive. Without both, the Renaissance would never have happened.

The fall of Constantinople in 1453 was one event among many that helped start the avalanche of social change that became the Renaissance. Many of the scholars fleeing Turkish depredations for the relative safety of the Italian city states were experts on ancient Greece and Rome. They took with them as many ancient scrolls and texts as they could carry or strap to the back of a mule. Many of these documents had never been seen before in Europe and Italian scholars eagerly devoured them.

One manuscript in particular that reached Florence around 1460 from Macedonia was a revelation. Marsilio Ficino – a priest, translator and scholar employed by grand duke Cosimo di Medici in Florence – was told to put aside his work on Plato and turn instead to this mysterious Macedonian manuscript. That he was told to stop translating the complete works of that revered ancient scholar Plato and turn to Hermes so Cosimo could read it before he died[8] is a measure of just how important the text was. As historian and Hermetic expert Frances Yates wrote: 'What a testimony this is to the mysterious reputation of the Thrice Great One.'[9]

The manuscript turned out to be an account of religion in ancient Egypt, one that deftly wove astral magic into its devotion to God. The author of the text was a legendary Egyptian sage called Hermes. Revered Fathers of the Church such as Lactantius and Augustine had mentioned Hermes in their writings and praised his piety and wisdom. Finding a manuscript he supposedly wrote was as close to a divine revelation as it was possible to get.

Hermes Trismegistus was reputedly an Egyptian sage whose work was said to pre-date that of Moses and therefore Plato as well. These historical figures were regarded as exemplars of learning and piety and were venerated by Renaissance scholars. However, greater respect was accorded to Hermes, who was supposed to be much older than these two pillars of Renaissance wisdom and as such held in much higher esteem. 'Trismegistus' means thrice-great and he was known by this title because he was a priest, philosopher, and law-giver or king.

The extreme age of the Hermetic writings (Ficino called his translation *Pimander* which was the name of the first book in the treatise) led Renaissance scholars to believe that they were reading the oldest, purest wisdom – as close to divine truth as it was possible to get. This was *Prisca theologia* undiluted by centuries of interpretation, debate and copying. It was an instruction manual for the creation of a better world. It didn't get much better for the learned men of the Renaissance.

The contents of the writings only served to cement the 1000 year reputation of Hermes. They may talk of spirits and magic but there is nothing diabolical about them. The texts are suffused with an extreme piety. They are concerned with opening man's eyes to the greatness of God rather than tempting him into evil. They contained an account of the creation of the world reminiscent of Genesis, and echoed a lot of the work done, apparently centuries later, by Pythagoras and Plato. Later parts of the text seem to prophesy the coming of the Son of God and the spread of Christianity.

The Hermetic texts reveal the methods men can use to gain salvation or obtain insights into the divine. It depicts the earth and everything material as being part of a 'world soul'. This immanent soul acted as an intermediary between the divine intellect and the material realm. God influenced events and material beings on earth via the world soul.

The parallel with astrology, in which the movements of stars affect events on earth, is obvious. However, astrologers only interpret what they see, making no attempt to influence events. They try and work out what will happen to someone as a result of competing astral influences rather than manipulate the situation to their own ends. Hermes, however, was advocating something much more radical. The Hermetic texts taught that by using astral magic, calling on the known sympathies between stars and events on earth, it was possible to change the course of the world. Instead of waiting for events to just happen, the magus should be able to steer earthly events clear of disaster or, if he chose, towards order and harmony. Using the Hermetic texts it should be possible to learn how to bend astral powers to the will of the magus.[10]

The writings of Hermes Trismegistus enjoyed enormous popularity and were hugely influential throughout the Renaissance. Every scholar was acquainted with them and many framed their researches using Hermes' theological and spiritual outlook.

One enthusiast, a Dominican monk called Giordano Bruno, was so struck by what he read in the Hermetic writings that he left his monastery and travelled round Europe preaching an alternative to Christianity based on the writings. He was so convinced Hermes was right that he, foolishly perhaps, travelled to Rome seeking an audience with the pope to tell him about the great and ancient prophet. Unsurprisingly he never got to see the pope. Instead he was tried as a heretic, imprisoned for eight years and then burnt at the stake at the Campo dei Fiori in Rome in 1600.[11] Others may not have gone as far or risked as much as Bruno but by this time the influence of the Hermetic texts was everywhere.

Copernicus cited Hermes in the introduction to his master work *On the Revolution of the Heavenly Spheres*. Medical pioneer William Harvey knew about Hermes, as did Robert Boyle and Isaac Newton. At least one unlucky child in Hampshire was even baptized with the name of Hermes Trismegistus.[12]

The Hermetic texts, however, were more than just a magical handbook. They also offered a fresh view of humanity. No longer were we subject to the iron rule of fate. Hermeticism freed mankind and left him to forge his own destiny for good or ill. It emphasized the dignity of man who was no longer born in servitude but had been given the means to free himself if he would but seize the chance. Hermeticism chimed with Renaissance humanism and gave mankind a nobility of purpose sorely lacking before.

The widespread adoption of the Hermetic/humanist philosophy was a turning point in mankind's relationship to nature. Now it could be made to bow down and serve man. No longer locked in servitude, mankind could take charge of his own life and test his own ideas against the world. He became more than just a thinker: he was a tinkerer, a trier, a scientist.

Putting the I in Science

The energetic experimentation that Hermeticism encouraged has been credited with starting the scientific revolution that is still rolling today. Certainly its emphasis on man as operator, drawing power from nature led many, including Dee, to begin their experiments instead of just thinking about them.

While keen experimenters are all around us today Hermeticism isn't. In the early seventeenth century careful study of the Hermetic texts revealed that they had actually been written in the first and second centuries AD. The prophecies they supposedly contained were no such thing; in fact they were written quite a long time after the events they purported to predict. The whole Hermetic project was founded on a huge dating error.

Thankfully the attitude it engendered remained and, freed from some of its religious and magical trappings, probably grew stronger as a result.

What is most important about this episode is why men were so ready, eager even, to accept the Hermetic story about the way the universe works, and why they were so keen to believe there was a reality above, beyond and between that which they saw every day. They accepted it because it fitted so well with their world-view. It seemed to offer a plausible response to the oldest of impulses that is always asking: Why are we here? What does it all mean? It helped end the quest for meaning and connection that we embark on from the moment we draw our first breath and only give up when we let go of our last.

Hermeticism gave Renaissance scholars a place in the cosmic scheme. It made them matter when, for centuries, they had been dwelling in the mud almost beneath notice, born into sin and condemned never to escape it. It gave life meaning. In some senses this is the least remarkable aspect of Hermeticism. Physics in the broadest sense of the subject has always been religiously inspired. This was never truer than during the Renaissance when the world was taken to be riddled with spiritual essences as much as it was made of hard, soft or sodden matter.

Magic and science were inseparable during the fifteenth and sixteenth centuries. If asked to describe the difference between the two a Renaissance scholar would be at a loss, because there was so little evidence to support any approach to a problem such as treating a stab wound. Who was to say why one solution would work better than another? To the Renaissance scholar who was part-seer part-scientist, it made as much sense to treat the weapon that caused a wound as it did to clean and bind the injury itself. Untested ideas about the 'sympathies' that existed between weapon and wound made this approach eminently sensible to the educated man of the times.

Nevertheless, Hermeticism was not popular just because of its affection for magic. John Dee and other fifteenth- and sixteenth-century scholars practised the Hermetic arts because they seemed to be a straight route back to a purer Christian time. By helping man recover his original state Hermeticism promised to leave him at one with the universe. However, this tendency to use ideas about the way the universe works to give man some spiritual ease is far older than Hermeticism. In a profound sense physics has always been about stilling the restless human spirit as much as it has been about working out how the world fits together.

Science For Starters

All science stems from the need to understand the world around us. Astrology and astronomy were the first sciences men practised because they instantly made the world tractable. They emerged when men realized they could link the turn of the seasons to the movement of the sun and planets[13] against the fixed starry background.

Even if you don't know the names of the constellations when you look into the night sky, certain shapes suggest themselves. Our visual system has barely changed in the 100,000 years or so that modern humans have been around. We have the same perceptual mechanisms as our most ancient ancestors and pick out the same shapes and clusters they did. We may explain the shapes in different ways but we still see the same ones. We look at the skies through ancient eyes.

Curiously cultures in different countries chose largely the same constellations when they were dividing up the sky. Basic perceptual mechanisms led men and women everywhere to see the same shapes in the sky, even if they gave them very different names. The Plough is familiar to most Westerners, though Americans prefer to call it the Big Dipper and the French know it as the Saucepan. The Skidi Pawnee Indians in North America saw it as a stretcher on which a sick man was being carried. The ancient Chinese thought that its shape suggested a chariot to carry the Celestial Bureaucrat and the ancient Greeks only saw it as part of the larger constellation Ursa Major, or the Great Bear.

However, the sky is not static. It swings around and the sun, moon, and other bodies move across the face of it in a regular pattern. This is a pattern that can help you plan your year and predict what is going to happen. It gives you control. Suddenly the world becomes manageable. Life is no longer a headlong rush, and you can plan, prepare, and create time to rest and think.

Certainly many of the constellations we recognize today are as old as civilization and must have been known and used long before that. Many of them got their names because the sun passes through them at a key point of the year. The watery constellations of Pisces, Aquarius and Capricorn (sea-goat) sit in the part of the sky where sun would have been during the Sumerian rainy season. Aquarius was important for the Egyptians too. The sun passed through it during July when the waters of the Nile had overflowed its banks and flooded the land. Taurus may have got its name because it rose in the evening during October. This was when cattle were coupled and the time when the waters of the Nile receded enough to let the land be ploughed.

It was a short conceptual step from noticing that the tides were related to the phases of the moon and that the swing of the sun through the zodiac traces the seasons to giving all celestial objects and events the ability to affect what happens on earth. Chinese astronomers thought that the first yearly appearance of Arcturus was the trigger for spring. Egyptians attributed the flooding of the Nile to the power of Sirius. The intellectual leap was as likely made because it was the only barely plausible explanation for such capricious events as earthquakes, floods and plagues.

It was an even shorter step to making celestial movements the pilot of the human soul and body. The Egyptians took this further than most and associated each planet with a particular part of the human body. They allotted the left eye to Saturn, the tongue to Mercury and the right nostril to Mars.

The relics of this early reliance on astrology are still with us today and are woven into our lives and language. The word 'desire' comes from the Latin word *desiderare* meaning without a star and an unfulfilled outcome.

Influence, influx and influenza all come from the Latin *influentia* meaning emanation of ethereal fluid from the heavens affecting mankind. We still describe each other with words from a time when the heavens were thought to determine our attitudes. Jovial, saturnine, and mercurial all describe the character of someone under the influence of those planets.

The more subtle movements of the heavens have also become embedded in our culture. It has long been known that the axis of the earth is slowly describing a circle in the sky 26,000 years in circumference. This gradual shifting is known as precession and means that the sun is gradually slipping back along the zodiac, moving back a degree every 72 years and a whole sign (a twelfth of the sky) every 2,160 years or so. This is why we will supposedly enter the Age of Aquarius sometime soon, as the sun is supposed to start rising in the sign of the water-carrier.

Precession has also been advanced as one of the reasons that Michelangelo sculpted Moses with horns. He was supposed to be alive at the time when the sun was just starting to rise in Aries. Some scholars dispute this explanation, however, and say that Michelangelo gave Moses horns because of a mistranslation. The original said that Moses had rays or light emanating from his head but this was wrongly translated as horns. But there are more certain signs. The Christian symbol of Jesus as a fish can also be linked to the fact that he was active just as the sun was rising in the new sign of Pisces.[14]

The science of astronomy/astrology reached its

fevered peak in ancient Greece and Rome. Pythagoras of Samos was among the first to try and take physics further into the spiritual realm with the aid of astronomy and astrology. Pythagoras was dazzled by the way that numbers could capture the movements of the planets in a simple form. A simple equation captures all you need to know about an ellipse drawn on paper or described by a planet moving across the sky. The short mathematical trick applies just as well to both.[15] This means that you can hold the secret to predicting the movements of the planets in your hand; you can write it on wax and carry it as a talisman. This is power indeed and it intoxicated Pythagoras.

The regular celestial cycles of four seasons, thirteen lunar months and 365 days only served to convince Pythagoras that number was the key to unlocking the secrets of the universe and the essence of reality. Pythagorean cosmology was as much a meditative tool as it was a practical science. Contemplation of the numerical relationships were supposed to help followers rise from the physical plane and contemplate the divine order at the centre of all.

It is no exaggeration to say that physicists have been trying to do the same thing ever since. An unbroken chain stretches from Pythagoras through the Renaissance scholars to modern physicists. All of them are trying to use physics to lay bare the universe, contemplate the harmony of reality, and calm the turmoil at our heart.

The spirit that spurred them to start their experiments unites them. It is the reason they started tinkering with the universe in the first place. The same urge made Dee seek out angels in Renaissance England and today makes particle physicists smash subatomic particles together and examine what's thrown clear of the wreckage. Einstein said: 'The longing to behold harmony is the source of the inexhaustible patience and perseverance with which [the physicist] devote[s] himself.'[16] They are all looking for the same thing – a deeper understanding of humanity and how we fit into the greater scheme of things. Particle physicists, Renaissance magicians, ancient Greeks and the builders of the henges are all as one in responding to it.

Modern physicists may be more bashful at declaring this than John Dee and his forebears but the urge is there nonetheless. The physicists' dream of obtaining dominion over nature, freeing our fellows from starvation and penury with the jewels of technology, thereby creating a Utopia where all go fulfilled and none want, is thinly disguised Christian theology. It is religion that has swapped its vestments for a lab coat. No more, no less. It also helps explain the rash of physics books with God in their title.

They are not confronting the universe, however, as much as they are confronting themselves. They are tackling the central problem of being human. They want to understand themselves and the events that befall them, their family and friends. They want to feel connected to the larger universe and perhaps gain some control over their lives.

You, too, will have confronted your humanity if you have ever wondered why we are here, what forces shape our lives, and what we would accept as an adequate answer to those questions. Self-reflection is the great curse and joy of being human. All those shamans, physicists, spirit guides, armchair philosophers and scientists are trying to do is understand how we fit in, to get at what it means to be human and give our lives some meaning. No one likes feeling helpless and alone. Sometimes it seems that in 5000 years, never mind 500, little progress has been made on framing an acceptable answer. Until now that is.

The History of Worrying

All those people embarked on their quest because, like you and me, they are worried. Worried that they, and the rest of us, are missing something. Humans are pattern spotters par excellence and it's because we're so good at it that I can sit here in a warm room listening to a CD while I write this on a desktop computer. Otherwise I'd be out hunting for food, competing for a mate and living in a tree.

Pattern spotting is a damn useful ability. If you don't make the connection between birds flying south and cold weather, you won't know when the herds are going to migrate and you'll be badly prepared for the harsh days of winter when food is scarce. If you never realize that lions rarely hunt alone, the next time you spot one sneaking through the veldt you won't think to look for the one creeping up behind you and probably won't live to see tomorrow.

This everyday empiricism can be a good guide but it is not infallible. Over the centuries it has given rise to some curious beliefs. Some are useful and definitely work. Thousands of lives have been saved in earthquake-prone regions in China by watching how animals behave. Dogs, cows and birds, as well as other animals, often become agitated or act oddly shortly before a big quake.

Other beliefs are understandable if wrong, such as the late-seventeenth-century belief that butter was yellow because cows ate a plant called yellow crowsfoot. Others, such as the belief that mice are born spontaneously from piles of dirty shirts, are just bizarre[17] though it doesn't take much imagination to work out how they arose.

This ability to spot patterns, to learn and make associations between separate events, serves all organisms well. Learning is a basic survival trait, but for us humans it has value beyond its ability to help us stay alive. It is useful for scientists too. Much of the edifice of modern science is built on predictability, observing a chain of consequences that can be relied upon to take place. For early scientists this might have meant simply noticing when the sun reached its most southerly position in the sky so they would know that spring was on its way. It might have led them to prepare some surprises for those returning herds. For Newton it meant that in 1666 when he saw the moon sailing between the trees and then an apple falling to earth, he could make the imaginative leap that links both and uses the same law to explain their motion.[18]

In more modern times it is even more useful. If, for

example, a young chemist had never noticed that *Staphylococcus aureus* bacteria didn't grow on a culture accidentally contaminated by a green mould called *Penicillium notatum*, then we wouldn't have antibiotics. An awful lot more of us would never have lived to see this summer.

The general success of our pattern-spotting talents has consequences for human psychology too. It has seduced us into believing that there is an identifiable sequence of events behind everything. As physicists racked up success after success in the early days of the science they, and we, have only become more convinced that everything works this way.

The success of Newtonian physics had a profound effect on the way that the universe was viewed. It gave birth to the notion of a clockwork universe in which everything was ticking round with regular pre-ordained precision. This led to the belief that perfect knowledge of the past would give perfect knowledge of the future. All it took was a mastery of the details and how they interacted. The grand design of God was becoming clear.

It was this that led Marquis Pierre Simon de Laplace to remark: 'We may regard the present state of the universe as the effect of its past and the cause of its future. An intellect which at any given moment knew all of the forces that animate nature and the mutual positions of the beings that compose it, if this intellect were vast enough to submit the data to analysis, could condense into a single formula the movement of the greatest bodies of the universe and that of the lightest atom; for such an intellect nothing could be uncertain and the future just like the past would be present before its eyes.'

Given that much of the world does follow a regular pattern, it is no surprise that we sometimes expect predictability and get none. It is easy to be seduced by our success and assume we can take this pattern spotting to its limit and explain everything. After all so much of our daily life is predictable. Every day the sun rises, rain falls, sugar dissolves in tea and one heartbeat follows another. So many things run in their regular groove that we take it all for granted and assume that everything must be like this. It is not unreasonable to think that we can take this predictive ability to the limit.

Following from Newton this is the promissory note physicists advance even today. Give us the time, money and computational power, they say, and we'll work out all the particles that are involved, how they can interact, wind everything back and forth in time, and spoil the story by telling you the ending. Reductionism, they say, is all you need to unravel this tangled knot of a world.

Sadly it looks like it is going to take a little more than that. Just because we can explain what might happen next in some situations does not mean we can do the same for every situation. There are some events, such as earthquakes, avalanches and stock market crashes that remain stubbornly unpredictable. Some phenomena, such as turbulence, are impossible to capture with existing mathematics – there is just too much going on. We can't model or even begin to understand what is happening.

Part of the problem is that if you reduce many of these systems to their constituent parts they lose their defining properties. For example, snowflakes have no in-built desire to do damage and studying one will tell you nothing about what motivates an avalanche, but put enough of them together and they can raze a town to its foundations. Stock traders do not want to lose all their money but the wrong sale at the wrong time can see them swapping a condo for a cardboard box. All too swiftly these events can teach us the limits of our learning and put us only a few steps ahead of our ancestral hunter-gatherers.

Similarly physics may be able to explain what was happening during the earliest moments after the big bang but it has proved very bad at explaining just how we are able to work this out. It has little to say about how human intelligence emerges or even how any intelligent life emerges. The events that physics is good at explaining are far removed from those that befall and confront us every day. Nevertheless, it should be realized that physicists have not left real life until last because it is easy; they have left it until last because explaining such events is really hard.

It is no error to say that although physics can tell you how to split an atom, it cannot yet explain why buses come in clumps or how those phantom traffic jams on motorways arise. Given that we see order all around but cannot see how it arises, it is no wonder that in the past we have ascribed its genesis to a higher power. It might be galling to admit but we are surrounded by order that physics cannot explain.

If you watch the way a flock of birds moves across the sky it often looks choreographed as they jink and turn to some hidden rhythm. Order emerges but no one is in charge. Look at the spread of veins in the back of your hand and then look at the spread of branches on a tree. There is similarity, order and regularity here, but until now no one can say how it comes about. Take a fern leaf and lay it on the page of an atlas showing the coast of Norway. They look worryingly similar. The patterns of frost on a window can share an uncanny resemblance to the cracking in the glaze on a doughnut. Similar structures confront us everywhere we turn. Fern leaves, fiords, and the branching found in lungs and leaves are regular patterns, not chaotic, but we have had no clue as to where this order might come from. Living organisms are the most obvious example of order at work and play. By the same token no one can explain how life has arisen, nor what sustains it, but we are here nonetheless.

It is clear from the similarities that fiords and ferns share that some ordering principle is at work. It is all around us, like the mist rising off a forest floor at dawn. Many theories have been advanced to try and explain what is going on here but all have been shown to be lacking. In the 1960s considerable excitement surrounded an idea dreamt up by French mathematician René Thom which he called *catastrophe theory*. This provides an elegant way of describing some aspects of turbulent fluid flow. Specifically it seems quite good at predicting when liquids change from one state to another – they undergo a catastrophic change. Hence the name.

Thom claimed that he had uncovered a general principle that could explain almost any complex system

– everything from embryological development to social revolutions. They could all be seen as fluid flows. Sadly the promise of the theory was not borne out. Thom's ideas are still used and are immensely helpful for mathematicians modelling fluids but they failed to be more widely applicable. The problem was that Thom's theory only applied to a few of the many complex systems in nature – the ones that break down into roiling turbulence. The rest, those that produce persistent structures such as you and me, it had little to say about.

So scientists have had to try again to find another explanation. Currently the only help physics is giving is in cataloguing all the situations where this concealed order, this hidden rhythm, is turning up. Ever more finely grained sensors are revealing that a vast range of phenomena have a common cadence. Everything from the words we use to the beat of our heart plays out to this tune. Yet the impetus for it all remains hidden. Stubbornly so.

Great chunks of our lives are playing themselves out to a frustratingly inexact rhythm, like a piece of poorly played music. It is one that we all recognize but we can't hum even if every day we prove that we know how it continues.

How many times have you had a hunch proved right? Have you ever felt that an event of some kind was looming and been curiously satisfied when it did happen, even if it caused you anguish at the same time? At some level you knew, even if you couldn't articulate it. To such expert pattern spotters as ourselves these hints are maddening. We find it annoying that we cannot extend our understanding to the areas that we hold more dear. They go some way to explaining why physicists feel the need to push their theories to the limit and make a little insight go a long, long way. Some would say too far.

Certainly the explanations that physicists have offered until now have done a poor job of easing the ache we feel or assuaging our urge to feel connected. There is too much left unsaid, too much order that they cannot explain.

On a subliminal level we know that the myriad tiny events of our day have a hidden tempo and it is our recognition of this that makes us suspicious. It makes us wonder just what is behind the fall of events. It turns some of us into physicists and some into mystics. Some search for an answer with drugs or the adrenalin rush of extreme sports. The rest of us just fret. We have an inkling that just beyond the scope of our sight, scent or hearing and the reach of our intellect is a profound insight that would explain everything in one shattering moment and leave us at peace.

Unfortunately it is not proving to be as simple as that. The secrets of our universe are proving hard to tease out. In John Dee's day the lack of explanations for so many commonplace events gave rise to beliefs in magic and a prevailing superstition. If you have no way of understanding what is going on, it is very easy to say that an invisible power is at work.

Although we have lost many of the superstitions that comforted our ancestors, we still keep others. Now, of course, they offer almost no comfort at all because much of the magic has been stripped from the world. We are not as ignorant as we used to be in the sixteenth century, but we are more scared than ever because we have so little faith at our back to hold us up. Equally so many of the explanations accepted by our forebears have been shown to be false that we can't honestly rely on them. We have swapped ignorance for unease and a rash of books that sell salvation for a sixpence.

The question thrown up by our restless intellect and the simple fact of being alive, of being an entrant in the human race, remains unanswered. Even religion is little more than a palliative. It puts off answering the question about what life is for until you are dead and can no longer tell anyone what the answer is. By that time even you are past caring. Religion demands faith, a taking on trust, that increasing numbers are loath to accept.

There is an ache, a deep-seated sense of curiosity allied with an inkling that we are missing something, that demands a response. Yet we do not know how to salve it, how to find our place and be at peace. Until now that is.

Finally after centuries of niggling doubts and half-right guesses we are close to understanding the forces that shape our lives and explain a huge swathe of the events we encounter every day. A common dynamic is thundering through the world. It is the engine powering how bodies are built, how they interact and the roll of history. The answer can be found within the pages of this book and grows out of work in a relatively obscure field of physics.

When philosophers talk of such things they call it metaphysics. The word comes from the Greek and means 'beyond' or 'after' physics. There are two reasons why it was given this name. Aristotle, the first philosopher to write a book on the subject, wrote it after he had written one called *Physics*. So because it came after 'physics' that was its name. The second reason is that it deals with subjects that are beyond the merely material, the everyday stuff that physics and physicists concern themselves with. It isn't about inclines and acceleration; it's about ethics, knowledge, destiny and freedom. It is concerned with questions such as: *Who am I?* rather than *How fast did it go?*

Aristotle's *Metaphysics* begins: 'By nature, all men long to know.'[19] By this he meant that we long to know why we are here, what our world is made of, and what our place is in it. Metaphysicians try to do this by seeking a fixed vantage point from which they can view and make sense of the world. They look for a starting point that is immune to changes in fashion or intellectual currents. They attempt to stand back from the rush of the human race and gain some perspective on what is going on and how it all fits together.

This problem of achieving some inner peace has become more urgent as we have discovered more about the universe. We urgently need a way to come to terms with the fact that we live in a quiet leafy suburb of a universe that rolls on and over us without a backward glance. Our inchoate yearnings demand an answer.

The history of humanity has been a story of gradual relegation to the sidelines. We have gone from starring role to sidelines, to just another planet, to a small planet in a quiet corner, to a dot that will come and go without anyone noticing, all in a few short millennia. We'll notice but it's not clear that our view counts.

Now we are cold and lonely creatures eager to communicate and connect, and we need a way to cope with the crisis. Strangely it is physicists, rather than philosophers, who are helping us understand and explain events that until now have failed to fall to our inquiries.

Existentialist philosophers say that this search is in vain, that there is nothing to find. They say that we are hollow within and will never find comfort because we are alone. Although this may seem a bleak prospect Jean Paul Sartre, the greatest exponent of this philosophy, says that instead we should see it as almost heroic, a noble calling. The job of every human is to conquer this solitude and reach an understanding with the world. As Sartre says in his masterly defence of the philosophy *Existentialism and Humanism*: 'Man is nothing else but that which he makes of himself.'[20]

For all Sartre's defence the existentialist world is still a bleak and lonely one. Worse, it does not seem to admit the existence of those inchoate yearnings that afflict us all. Even if they are illusory – sparks in the darkness – we can all see them and follow their flight. The ache remains and the call goes unanswered.

Well, unanswered until now that is. The answer is at hand, and the waiting is over. The response and the explanation can be found not within but beyond, down by the water. On the beach.

Chapter Two

The Littoral Truth

The behaviour of large and complex aggregates of elementary particles, it turns out, is not to be understood in terms of a simple extrapolation of the properties of a few particles. Instead, at each level of complexity entirely new properties appear, and the understanding of the new behaviours requires research which I think is as fundamental as any other.

More Is Different, Peter Anderson

To see a World in a Grain of Sand
And a Heaven in a Wild Flower,
Hold Infinity in the palm of your hand
And Eternity in an hour.

'Auguries of Innocence', William Blake

Son of Beach

It is 19 July 1999. Summer has conquered Catalonia, particularly the playa at Castelldefells, a small Spanish town lying on the coast around five miles west of Barcelona.

It is a glorious day, with not a cloud to be seen in the blue, blue sky. Just the sun beating down on those lucky folks who can spend the day at the beach – lucky folks such as my girlfriend and I. We've decided not to waste the day sightseeing or shopping. Instead we are doing something much more worthwhile: sunbathing.

I'm sitting beneath a beach umbrella, contesting the skewed square of shade it casts with Clare, who is stretched out on a beach towel next to me. We may be enjoying ourselves now but the heat of the day and the hot, hot sand made a trial of the walk from the road to the spot we picked for our sunbathing. During the trek across the sand our flip-flops kept falling off or filling with scorching sand that hobbled our progress. We were burdened with parasol, towels, water bottles, books and sun lotion too, all this to enjoy a couple of hours in the sun.

We've settled on a small rise landward of the high-tide mark from which the beach shelves down and away to the sea. Now that we've pitched our parasol and laid out our towels we realise the trek was worthwhile. The gentle onshore breeze blows strongest over the knoll we've picked and gives a measure of relief from the beating heat of the day.

Although the wind helps keep the temperature down, this Spanish summer is really a bit too much for us pale-skinned Brits. But we don't want to lose face by leaving too soon. We tacitly decide to stick it out for a couple of hours to show how cosmopolitan we are. To keep ourselves within the bounds of the cool pool of shadow beneath the umbrella we're forced to keep up a slow shuffle as the shadow slowly ticks around the pole of the parasol in time with the sun. If we stay put we'll gradually become exposed to the cruel heat of the sun.

Clare is dozing in the shade and I'm sitting sipping water, listening to the bottle hoot agreeably every now and then as the wind gusts across its open neck. I've bravely got my legs from the knee down in the sun. I'm resting my feet on my upturned flip-flops because the sand is too hot to touch for anything more than a moment. Even behind my sunglasses I'm squinting to cut down the glare reflecting off the sand and sea.

Suddenly the oven-like heat overwhelms me. It feels like it's sloshing through me like a hot heavy wave. Despite the water I'm sipping, the breeze and the shade, it's just too hot. I need to cool down quickly, so I stand up and head for the Med. Clare takes this opportunity to scoot over and claim more towel space.

At first the shock of the cold water makes me gasp but as I wade deeper it's the strength of the swell that takes my breath away. The beach shelves away swiftly beneath the waves and before I know it I'm up to my thighs with the surf smashing against my legs. It's hard to stand and for a while I'm left struggling to keep my feet, let alone wade deeper. I stagger as I'm caught between one heavy wave trying to swat me aside and another behind it which slaps across me from knee to nose. Suddenly I'm soaked, left spluttering and laughing from the short sharp shock of the cold water. At least I'm cooler now.

Here where the sea is shorts deep, waves wash back from the beach and win fresh energy from those yet to founder. The rhythm of wave, wash and counter-wave is regular but what happens once a wave has broken is completely unpredictable. In the surf zone there is no pattern to the returning roll of the waves, just a white shattered rush of roiling water. The cross-currents make it difficult to walk and many people stumble as they pass through it, dodging one large wave only to be caught by a bigger one beyond or another wave pushed into them by the crush of water.

I wade beyond the surf zone, out to where the water is chest high, where the breaking waves are rolling in, gathering their weight before they lift, rush forward and break upon the sand. Here where the swell rises and falls there is a region of calm before the crash. Waves that will drop heavily on the beach are mere swells here and buoy me up and down as they pass by.

I rise and fall with the waves with little fear of being overwhelmed. I'm not entirely free of trouble because some really big waves can't wait for the beach and break even here. If they catch me unprepared I'm in for a ducking. For the moment though I can happily bob up and down in water that is just the right side of cold. Chilly currents occasionally pass by but on the whole the temperature is refreshingly right.

Bathing Beauty

In the troughs between the waves I can just about touch the sandy sea floor with my toes. I'm stable enough to watch the narrow strip of beach before me and the people upon it. From here it looks like a disaster area with bodies strewn everywhere. Near naked people are laid out for miles on end, all the way from the marina a mile or so away to my left, past and beyond me down the coast as far as I can see. The sun-worshippers lie as still as if they had been anaesthetized.

Anyone looking from above at the way people are distributed along this narrow strip of sand would probably see an eerily regular pattern. The distance between the groups and individuals spread along the strand is remarkably uniform. No one is too close and gaps left by departing sunbathers are swiftly filled. People trudge for minutes until they find a spot close but not too close to other groups or lone bathers. The only anomaly is probably the area around us stand-offish Brits who'd rather not get too close to other sun-worshippers.

The forces at work here – national mores, social psychology and limited space – meet and mesh to produce something strikingly regular. Not regular like bricks are regular but regular in the sense of orderly in the same way that clouds are. Some hidden principle gives clouds their characteristic tuft-upon-tuft-upon-tuft shape. It makes them instantly recognizable, even if no one can write down the mathematics to explain what is going on. The same is true of sunbathers. There are forces at work that everyone understands and shares but no one can articulate. You just know if you are too close to another group just as you know if you are too hot or cold.

As I watch young and old, male and female alike, pass in and out of the water, I wonder if the way they approach and enter the sea says anything about them. Is their larger personality revealed by whether they are waders, splashers or paddlers? Certainly people deal with the waves in very different ways. Some stand on the margins happy to have nothing but their toes washed. Others advance cautiously. They only move into deeper water when they are sure of their footing and have got thoroughly used to the cold and they shriek when a wave gets them too damp too quickly.

Every movement of the careful waders betrays their unease. They know that if they stay too long in one place the retreating water will take the sand beneath their feet with it and they will be left in a worse position than before. It makes me wonder if they are this careful with every decision they make, only advancing when they know what is at stake. With their affections perhaps? Others, solid middle-aged citizens all, find what they can cope with and stay put, moving around only enough to keep their footing. Some, mainly the young, throw themselves through every wave creating as big a splash as possible. A few are bested by the big waves which fling them back towards the beach and these few withdraw for a while before trying again, and again, and again. Older folks, who perhaps know the beach and waves better, make their way to deeper water beyond the waves with an admirable aplomb. If a wave catches them they philosophically shrug before keeping on. They let the water take their weight rather than constantly trying their strength against the crashing surf. Perhaps the wisdom with which they approach the rest of life is used here too. Perhaps all of human life is writ small here.

Now I'm getting too cold and paranoid about shark attacks so I half-swim, half-walk back to Clare and the heat of the beach.

Physics is at work everywhere here: the whole beach is a living laboratory. The waves breaking on the sand are the end point of a fantastically complex interplay of forces that are at work beyond the beach, even beyond the ships sailing along the horizon. The shelving beach serves to concentrate the energy pushing the waves on until they collapse under their weight. It is impossible to see where the impetus for any particular wave began, what coincidence of forces started it off. It is just as hard to predict the height of the next wave.

What is happening here is not confused and random. There are regularities to be seen in many of the events I witness. Gulls in flocks do not bump into each other or argue about the direction they should go. As they criss-cross the beach looking for food their turns are smoothly coordinated. They ripple like a living flag as they turn across the wind.

Sunbathers are not piled one on top of the other, nor thrown randomly about the beach; they seem to be carefully placed. The bifurcating patterns of the fronds of seaweed washing back and forth in the swell resemble land plants in the way that they grow. All of these phenomena, and many more, show that order, or an ordering principle, is at work.

Every event on the beach that I can see is an experiment in action. Ones that before now no physicist would have the courage, or methods, to tackle. With soaring gulls, sand sculpted by the sea, growing seaweed, Spanish sunbathers and breaking waves it is hard to know which influences matter most and how to go about plotting their interaction. There would be just too much information to sift through, too many influences to sort. The complexity and weight of the data would overwhelm a physicist keen to model them like a tsunami.

This frustrating inability to say anything meaningful about how many everyday events come to pass might be about to change. Of late a series of insights are convincing some scientists that what is happening on the beach, and many other humdrum walks of life, can be made sense of. Through the seeming confusion the dynamics of all these events can be teased out and understood. Even better it doesn't take immense computer power to tease out the forces at work or gain an insight into how such events arise. You don't need drugs or divine wisdom. You need only the humblest of tools to tackle these problems. Everything you need is right here. On the beach.

Son of Sand

People go to the beach for many reasons: to escape the seemingly senseless rush of life, or to get some distance from the pattern of events that often only seems rich in the ways it finds to thwart and frustrate us. But while they are relaxing in the sun they may be surprised to know that they have on hand, literally if they are lying down, everything they need to change their perspective and realize that life makes more sense than they think. Science at school was never this enjoyable, nor this straightforward. However, one thing remains the same on both beach and lab bench: you still have to pay attention.

On any sandy beach you will find all the materials required to understand some of the most pressing questions in science. It might help explain how life arose from a soup of dead chemicals, the unpredictable dynamics of earthquakes, the distribution of galaxies, the booms and busts of national economies, the drivers of evolutionary diversity, even why buses come in threes. It might even help to explain why it is only humans that seem to spend any time worrying about these things.

There is only one spot that will do. You should stand at the water's edge, where giddy waves throw cool water over your toes and then beat a hasty retreat with the sand they scoop from beneath your heels. Look down and you will see what you need right there at your feet. All it takes is sand, water and waves.

Of the three, sand is initially the most useful when it

comes to modelling and explaining the forces at work within populations, people and pulsars. You could be forgiven for thinking that sand is good for modelling nothing but castles and even then only when it is wetted enough to ensure the grains stick together while you are stacking and shaping them.

In Bankass, Mali, and Shibam in the Yemen people have exploited this stickiness to build, respectively, a mosque and an entire settlement. The town of Shibam is a magnificent construction, with sand and mud walls studded and strengthened with logs and poles. It is a sandcastle fit for a king. The mosque too has an earthy grandeur, with towers similar to those found on buildings designed by that crazy Catalonian architect Antonio Gaudí. Both are magnificent examples of what sand can be made to achieve.

But, as the builders of Shibam will tell you, if you use too little water then your construction – even if it is on a much more modest scale than the mosque in Mali – will fall apart just when you are putting the finishing touches to your finials and machicolations. They will also tell you that you shouldn't add too much water either or your sandcastle will never rise to be anything other than a heap. Washing machines and modern plumbing have caused havoc in Shibam.

Nevertheless, a heap can still have its uses. When it comes to explaining the forces at work within so many events and organisms, this heapability is the most useful property of all. The tendency of sand to collapse into a pile rather than stand up for itself has piqued the interest of a small group of physicists. They think that they can use this property to capture the qualities of the complex systems that until now have largely defied attempts to study them.

Pick up a handful of dry sand, make a fist around it, and then slowly let it slip out of the tube made by your fingers onto the beach. Your hand should be about 20–30 centimetres above the pile you are creating. Keep adding fistfuls in this way and you will quickly create a small shallow-sided pile. Add the sand slowly as the heap gets higher. Eventually the weight of grains falling onto it and gathering at its peak will become too heavy for the straining grains below and a tiny avalanche will sweep down the slopes of your pile.

Some of the sand you added will slip to the base of the pile leaving it larger but the sides will still probably be at an angle of roughly 34° with respect to the horizontal. If you had measured it earlier this would have been about the same angle as the smaller pile. Keep adding handfuls and watching avalanches of all sizes sweep down its sides. It is important to pause every time an avalanche strikes and let the sand settle down again before you add any more grains.

Gradually a pattern should emerge. Avalanches of different sizes keep occurring and returning the pile to its steady, shallow-sided, state. The avalanches will be of different magnitudes. Some will be huge and make it look like the whole pile is collapsing. Others will be localized small sweeps that stutter and stop before they carry away too many grains. If you watch carefully you will see that small avalanches are more common than large ones but all sizes should be seen if you watch long enough and add enough sand.

Curiously there is no relation between the amount of sand added and the size of avalanche that follows. Sometimes it takes a whole handful to set a side slipping, but if you add the sand more carefully a huge sweep of sand can be triggered by the bumping fall of a single grain.

If you called in a physicist to describe what was happening he or she would say that as you are adding sand to the pile you are driving it away from its steady state towards a critical point in its development. The steady state is the static shallow-sided pile and the critical point is reached when an avalanche sweeps across it. The spiky grains tumbling from your fist perch one upon another until it reaches the critical state. At this moment the weight of sand being supported gets too great for the grains it is resting on, triggering a collapse. However, it is hard to find out just what takes the pile to this point.

You can experiment and add grains one at a time if you want or in a steady trickle. It does not matter. As soon as the critical point is reached, change – in the form of an avalanche of sand – sweeps through the system. In this context the word 'system' is shorthand for whatever you are studying. The only caveat is that there should be competing influences at work within it. A waterwheel is a system, as is the weather. And so are you.

More than Moments

The systems physicists have in mind are those that cause big problems when they do something unexpected. When they do, stock markets crash, avalanches sweep down mountains, hearts stop and fault lines slip. Getting a better idea of the internal dynamics of what causes these events would be a boon to many fields. It might also save a few lives.

What heart muscles, stock markets, earthquakes and avalanches have in common is that they are made up of thousands, if not millions, of interacting units. A stock market is an outgrowth of the activities of thousands of traders shouting into as many phones or clicking as many mouse buttons. An earthquake is a summation of smaller slips in fault lines that riddle the crust of the planet. Get enough of these elements on the move and the effects can be felt around the world.

The outcome of all this interaction often appears to be a single event. A ripple spreading across the heart is felt as a pulse beat, the jostling of tectonic plates is captured as a number on the Richter scale and a stock market meltdown is measured by a single figure for how far the index fell. In truth what we see or feel is really the result of a much more complex underlying process. An earthquake is really a collection of many events on different scales as faults in plate boundaries slip, snag and slide to varying degrees. The buying and selling of shares by thousands of traders responding to a government policy change causes a stock market to fall and heart failure always has a history of smaller events leading up to it.

No single measure does justice to what is going on when an earthquake or a heart attack strikes, and this is the problem facing any scientist wishing to study them. What they have to do is find a way to model these systems without losing any of the subtle interactions between the elements. This is key because small changes can be very important in these complex dynamic systems.

These systems are a curious mixture of sensitive and sturdy. They are sturdy because they can soak up huge amounts of external input and return to a steady state that keeps chugging along. At the same time they are also exquisitely sensitive to small changes. The wrong change at the right moment and suddenly a mountain is on the move. Yet once this change has run its course, when the change has swept through the system, it will be back in a state of poise.

It used to be thought that to capture such complicated dynamics would take theories of an equally complex nature. Only with huge computers, vitally detailed simulations and aeons of time to run the programs would you get close to tracking what triggers change. However, it turns out that the opposite is true: with a very simple system you can mimic the subtleties of behaviour lurking behind many real-world events.

But, you might say, this is just sand. What possible relevance can this have to such vastly different phenomena as stock markets, earthquakes and the spread of disease? Well, while physicists can say little that is profound about the dynamics of earthquake tremors or stock market crashes, they can say something meaningful about systems passing through critical points. It turns out that the two are intimately related.

There are many real-world examples of systems with critical points. Adding energy to a system, often in the form of heat, tends to drive it to a new state. Often the change takes place swiftly once the critical point has been reached. Many of these systems have been extensively studied. Some physicists have won Nobel Prizes for work that contributes to our understanding of them.

Matter in Motion

This is a field with a long history. In 1869 the Belfast-born physicist Thomas Andrews published some of the earliest work on the critical points of gases. He did most of his work on carbon dioxide and established that there were temperatures above which it became impossible to liquefy gases by pressure alone. Andrews noticed that at low temperatures Boyle's law no longer held true. The law states that as the pressure increases the volume of a gas decreases. This is no longer true at the so-called critical point when it was hovering between its liquid and gaseous states. Van der Waals followed up his work with a more detailed description of the liquid–gas critical point.

An even better known system with a critical point is that familiar substance H_2O. If you heat water in a pan at sea level it will happily get hotter until it reaches its boiling point of 100°C – no more, no less – when the liquid water becomes a gas. It turns into steam. Get higher above sea level and these values change. On top of Everest, an altitude of around 10,000 metres, the boiling point of water falls to 72°C.

Among physicists this change is known as a first-order phase transition. Only at this temperature does a liquid become a gas. As it changes from one state of organization to another, from liquid to gas, it effectively becomes a different material with very different properties. In the case of water it is much less dense, a given amount of steam having 1600 times the volume of the same mass of liquid water.

At this critical juncture a tiny change, such as a change in temperature, might mean the whole causes the liquid water to flip swiftly into its gaseous state. The water is either one thing or the other. What we know about water in its liquid form is of little use in describing what is going on at this molecular interface.

The temperature at which water boils rises if you increase the pressure. As the pressure increases the change in its density is smaller. If you keep winding up the pressure the boiling temperature continues to rise and the densities of the liquid and gaseous phases start to converge. At a pressure of 218 atmospheres and 374°C the surface tension between the two phases disappears. The liquid water and the steam become indistinguishable. At this juncture suspended drops and bubbles of water scatter any light falling on it making the whole look opalescent and milky. It is quite beautiful. This is the critical point.[1] It is at this juncture that the two possible phases of water coalesce into a single phase. It enters a limbo-like region where it is neither one thing nor the other.

Many other substances, both gases and liquids, such as alcohol, ether, carbon dioxide and oxygen possess critical points. But there are better examples of systems that have a critical point and undergo equally striking phase transitions. One of the most widely quoted and studied is the iron bar magnet[2] that also has a critical point where its properties change abruptly. This critical point is known as the Curie point, which gives you some idea about how long it has been known (since 1895) and who was doing the early work on it.

As every school child knows you can use a strong bar magnet to pick up drawing pins or thumbtacks and other small metallic objects. Yet, if you then heat the magnet, after a while the pins will start to fall off as the iron begins to lose its magnetism. As the magnet gets hotter and hotter the pins will fall off at an increasing rate. Beyond around 770°C, the critical temperature, the bar is completely non-magnetic and all the pins will have fallen off.

It is useful to see why this happens. On a microscopic scale a bar magnet is made up of many millions of tiny magnetic grains. Like all magnets these tiny grains have very definite ideas about how they should be organized. They prefer to be arranged with opposing poles nestling against each other. They are just like any two larger magnets that, again as any schoolchild will tell you if asked, fiercely resist having like poles pushed against each other.

Thermodynamics says the same but for more complex reasons. It says that the miniature magnets will tend to assume the organization that needs the least energy to maintain. This is simple thermodynamics. If it takes less energy to keep magnetic fields aligned than it does to keep them opposed, the whole bar becomes magnetic. Which end of a magnet is north and which south depends on the history of the magnet and the orientation the grains decided on last time it was heated and cooled. As a result at room temperature the microscopic grains are highly organized and their magnetic fields reinforce each other. It is this co-operative bent that makes the whole iron bar magnetic.

As the iron is heated, however, the thermal energy being added to this system shakes up this neat organization. The grains are jostled and don't know which way to turn. Around the critical temperature of 770°C groups of grains may manage to organize themselves briefly but heat and the fluctuating magnetic fields of neighbouring large and small clumps will quickly destroy this island of stability. Clouds of magnetism will ripple up and down the bar, flickering in and out of existence. This is why the drawing pins fall off as the bar is heated. Only when the iron cools will the microscopic grains organize themselves again and render the whole magnetic.

Some astonishing work in physics has been done to explain the dynamics of the critical point and insights have been gained into what is happening at this moment as well as into a vast range of other, related, phenomena.

Critical Theory

Until the early 1960s physicists of all stripes used an approach called mean field theory to explain and/or predict what was happening in liquids, solids and gases undergoing change. Mean field theory is an outgrowth of a branch of physics known as statistical mechanics. It got its start as scientists were struggling to find ways to capture what happened to gases and liquids when they were being heated or subjected to different pressures.

As the name of this branch of science suggests it uses statistics to get to grips with phenomena that would otherwise be impossible to study. Up to a point it is good at predicting what will happen to systems composed of many 'bits' such as gases or liquids, which are made up of many millions of molecules. Statistical mechanics gives physicists a framework for dealing with the general properties of a complex system. It is useful in situations when they don't have to worry about what individual atoms are getting up to. It is a great tool for understanding the bulk properties of matter, which are often the only ones that are important.

There are around 10^{23} or so interacting molecules in a litre of hydrogen gas – far, far too many to deal with on a one-to-one basis. There is no way on earth, or off it, that scientists could track each one and record what happened when the hydrogen gas was being heated or cooled. Thankfully there is usually no need to track the trajectory of each molecule, or its collisions with its fellows or the sides of a test tube. What works just as well is treating the gas as a collective entity. Instead of trying to watch each atom and then use those observations to explain what is happening to the gas at large, statistical mechanics considers the big picture. It considers the average behaviour of large numbers of the same sort of particles (e.g. hydrogen molecules). It says that you don't have to watch each one to understand what will happen when the hydrogen gas is heated. The particles may be reacting in as many different ways as there are particles but that doesn't matter. All that matters is the average of all their behaviour. You don't have to know what will happen to all of those atoms if you can depend on the sum of all that activity producing the same result every time. A few rogue atoms with more or less energy won't mess up your calculations of what the average number of particles will be doing.

James Maxwell and Ludwig Boltzmann did the pioneering work on statistical mechanics and helped establish its fundamental principles. During their day the field was better known as thermodynamics. Maxwell's work on the kinetic theory of gases produced a statistical law that reveals the different velocities of the molecules in a gas. Using the law he showed how the collective behaviour of all those molecules determines the properties of a gas, such as its pressure and temperature. Boltzmann partly qualified this law by developing the Boltzmann distribution, an equation expressing the distribution of energies for a collection of particles at a fixed temperature. This may sound complicated but it isn't. It defines the rate at which temperature makes interesting things happen. It explains why the speed of some chemical reactions increases at higher temperatures. One of the formulae defining this field is carved on Boltzmann's tombstone.[3]

Mean field theory helps physicists delve deeper into the properties of those systems that statistical mechanics was helping scientists to catalogue. Like its forerunner statistical mechanics, as well as many other theories, it was the outgrowth of years of work by many different scientists. French physicist Pierre Weiss developed Curie's work on critical points and was the first to find them in ferromagnets. The work was taken on by Russian scientist and Nobel Prize winner Lev Davidovitch Landau, who in 1937 drew up the version of mean field theory widely used in the early decades of the twentieth century. The theory that he espoused assumed that the only fluctuations or changes that matter are those that take place on the atomic scale. It assumed that, in a ferromagnet, each microscopic magnetic grain was influenced only by the average behaviour of all the others and each minuscule magnetic field was thought to interact equally with all the others.[4]

A Moment of Clarity

There was every reason to expect that the success of statistical mechanics would be repeated when it was extended to critical points in other materials such as ferromagnets. It did an excellent job of predicting what happens to gases undergoing temperature and pressure changes and would do the same when it was extended to ferromagnets and such like. It was thought that it would be equally applicable to any other system going through a second-order phase transition. Well, that's what everyone hoped.

Landau, and many others that followed his work, thought that there was a linear relationship between the temperature of the magnet and the number of drawing pins that would stay attached to it. Surely, they thought, anyone drawing a graph plotting the number of pins held on a magnet against temperature would see a straight line – as the magnet got hotter pins should drop off regularly until there were none left. Except that it doesn't happen like that at all.

The transition from magnetic to non-magnetic in a heated iron bar magnet is quite abrupt. All of a sudden the ferromagnet just seems to give up. There is none of the smooth change that might be expected given the earlier experience with gases and the success of statistical mechanics in predicting what happens to them as they are heated.

This disagreement between theory and reality is easily revealed. One of the best tests of a theory is if it can make accurate predictions about the behaviour of the systems that it tries to describe before the experiments are done. Sadly mean field theory fails this elementary test. Mean field theory turns out to be unable to predict critical temperatures for many materials to within about 40 per cent of their actual value.[5]

For physicists this was deeply worrying. They were used to relying on the thermodynamic properties of a material being macroscopic, i.e. the change in the properties of a material should be proportional to how much you heat it. The averaged actions of all those tiny molecules should be enough to explain what was happening for the whole. The rule held largely true for gases. Yet with a magnet approaching its Curie point something was going on at a microscopic level that was managing to upset, however briefly, these macroscopic thermodynamic properties. Mean field theory was missing something, and something quite profound. But what?

For a good few years no one knew. All they did know was that mean field theory was rapidly losing friends. In 1944 Lars Onsager did some work on the thermodynamics of an idealized one-dimensional magnet called an Ising model. The results he obtained violated the predictions of mean field theory and started the debate about the shortcomings of the approach. In the late 1950s and early 1960s the limitations of mean field theory became even more apparent thanks to the work of scientists such as Cyril Domb, Michael Fisher, Sasha Voronel, Brian Pippard, Mel Green, and others. All of these researchers were busily cataloguing the places it failed to predict what actually happened.[6] There was a desperate need for a fresh way of explaining just what was going on during phase transitions. Scientists, like nature, abhor a vacuum and independently three groups of researchers found the key to a new way of thinking about what was happening during phase transitions.

One of the first to make sense of what was happening was American scientist Leo Kadanoff, who was at that time studying at Chicago University. In the early 1960s he was attracted to the study of second-order phase transitions because few people had done much work on them and like any eager graduate student he was keen to make his mark. It seemed to be a field that promised rich rewards for new explorers. He was intrigued by just what was happening at those fleeting critical points. Either side of them the behaviour of a system was perfectly understandable but at that point something was going awry. As Kadanoff says: 'Here was a tiny little world, just waiting to be explored and perhaps even captured'.[7] Kadanoff was keen to plant his flag.

The problem with mean field theory lies in its generality. Like its bigger brother, statistical mechanics, it deals with average behaviour. It insists that the behaviour of one particular particle is influenced only by the average behaviour of all others. It does not take into account the possibility that one particle can influence any other or that there can be relationships, correlations or a history of interactions between particles. It was obvious from messing about with a sand pile that a single grain can have a huge effect. If it falls at the right moment it can trigger a sweeping change that affects the whole system. One grain can take the rest of the sand with it. Mean field theory utterly fails to capture this possibility.

Eppur si muove

In the Christmas week of 1965 Kadanoff had 'a sudden vision' that showed him exactly how this could happen. A large bar magnet is made of many millions of microscopic magnets. At room temperature almost all the north poles of the mini-magnets are pointing one way and nearly all the south poles the other way. This is what makes the entire bar magnetic too. When the ferromagnet is heated the energy being added upsets this neat arrangement. Some of the tiny domains flip and the confusion of magnetic fields means that gradually the bar loses its ability to pick up pins, tacks and other metal objects.

Mean field theory can start to explain what is happening because initially there is a linear relationship between the gradual loss of magnetism and the increasing heat. However, as more energy is added this relationship starts to break down. As the bar is heated more and more of the mini-magnets begin to vacillate between different orientations, flipping now one way and then another. The magnetic pull of their neighbouring grains might influence them too. They might be flipped around or they might stay as they are. As the heat of the magnet rises this indecision becomes more pronounced.

In a magnet held at the critical point energy is being added all the time, so it becomes easier and easier for those miniature magnets to change their mind about which way they are pointing. The heat being added may kick them head over heels. There is always enough energy available to change orientation, so they have no reason to stay fixed and anything can happen. Orientation, and organization, can appear and disappear in an instant.

For this reason it can take very little for one magnetic domain to take a lot of the others with it. As a result correlations between the tiny domains can have an enormous effect on the properties of the entire bar. The orientation of one domain, an atom within it, or a few neighbouring domains, can convince all the others throughout the system to orient themselves the same way.

A cloud of magnetism can form, grow and swiftly ripple down the length of the bar. As it does so some regions within that cloud may suddenly flip, making the large domain shrink again. The result is a magnet crippled by indecision with larger and smaller domains of magnetism bursting into life and rippling along its length.[8] It is a system struggling to reassert order, yet is poised on the cusp of chaos. It sits on the dividing line between the two. Says Kadanoff: 'This process of choosing produces a beautiful, closed world with its own rules and its own internally consistent explanations.'[9]

Weighing the Scales

To characterize what was happening Kadanoff realized that the scale being used to measure what was going on was irrelevant. The changes in one tiny corner of the hot magnet could have their effects felt anywhere and everywhere. They weren't localized. There was no cut-off point beyond which the effects of one miniature magnet flipping would not be felt. The ripples could reach everywhere in the system, and wherever a ripple reached its effect on neighbouring grains would be seen and perhaps persist.

To make matters worse no matter where you looked

on the magnet the same push-me-pull-you interactivity could be seen. At the critical point organization could effectively spread over infinite distances. This meant that there was no distinctive scale you could pick to make sense of what was going on. It was all one seething interacting mass. If you took a square centimetre of the bar magnet, plotted the pullulating movements of the magnetic clouds within it and compared that with a larger or smaller section you would see the same degree of busy blooming confusion. If you took a peek you would see that magnetic clouds were forming, flickering, blooming and collapsing in the same way whether you were looking at the magnet as a whole, considering only a few neighbouring domains, or observing any size chunk in between. It doesn't matter. Within the large magnetic clouds would be smaller clouds with the opposite polarity forming and disappearing too.[10] The dynamics of one region, no matter how big or small, was characteristic of the whole.

Initially Kadanoff's insight was only a phenomenological one. It helped people think about what was going on. The experiments and mathematics had yet to be done to prove that it tallied with reality. Before that work was completed Kadanoff found out that several other scientists were thinking along the same lines. Different groups had chosen to characterize what was happening in the magnet in the same scale-free way. Separately Ben Widom and Michael Fisher at Cornell University were coming to the same conclusions. So were the Russian scientists Aleksandr Z. Patashinskii and Valerii L. Pokrovskii.[11] The work of the Soviet scientists remained hidden behind the Iron Curtain and only later got the recognition it deserved.

Around the same time that Kadanoff was experiencing his 'vision' the US National Bureau of Standards was hosting a conference on phase transitions and critical points organized by Mel Green. The conference brought together many of the scientists working in the field and served to concentrate thinking. The avalanche of work that followed the conference confirmed that Kadanoff was right to approach the problem in this way.

Kadanoff himself was not idle and, working with colleagues, he set out to determine whether his insight fitted the facts. He and his colleagues dug out and tracked down all the experimental and theoretical data about critical phenomena to see if what was reported could be explained by his ideas. They were relieved to find that his insight was confirmed and a paper saying as much was published in April 1967.[12]

It was about this time that the word 'Universality' first started to be used to describe this work on phase transitions. Kadanoff says the first time that he heard it mentioned in connection with the field was in a conversation with Sasha Polyakov and Sasha Migdal in a dollar bar in Moscow.[13] When it was first used many people knew that phase transitions were common but it has taken a couple of decades of work to reveal just how appropriate that name is. The word was used because it was becoming obvious that at the critical point the details of the system you were studying became irrelevant. The particles making up your system, be they water molecules or miniature magnetic grains, no longer mattered. What was important was the organization they displayed, the flickering hints revealed as it hung between order and chaos. As the experimental work got under way and data started to be collected the term became increasingly apt, but it was not until much later that they realized just how correct it was.

What was apparent from this flurry of work in the latter half of the 1960s was that to understand what was happening at the critical point physicists had to drop the notion of a single scale. The study of such things as the physics of radio waves, atoms and nuclei can be done using one characteristic scale; nothing more is needed. At the critical point this is not the case. Physicists had to stop looking for a single parameter that they could use to measure how well organized, or otherwise, the magnet was.

Unlike statistical mechanics there was not just one level they could use to measure the system under study. They had to realize that interactions in the bar magnet could involve a couple, some, all or none of its constituent parts. An understanding of what was happening had to take this possibility into account. Any arbitrarily chosen unit of measurement would at once be too fine and too coarse to capture the organization in the system. Order could spread throughout the system, persist for a surprisingly long time, and make its effects felt far from its starting point. You could miss it if you had your dividers set too close or too far apart.

Wilson Wins

What was needed was a scale-free method of describing what was happening. This mathematical explanation had to be able to capture the fact that the same dynamics were at work across all scales, everything from the macroscopic (the whole bar magnet) to the microscopic (the atoms within it). Perhaps unsurprisingly, finding a way to measure what is happening by throwing away your metaphorical yardstick was tricky. In the event the man who managed the feat got a Nobel Prize.

Thankfully physics already had some experience of dealing with inexact, in this case nearly infinite, quantities. There were established techniques that physicists could draw on to help them gain some idea of distance without, in effect, using a tape measure. In the 1940s quantum theorists such as Richard Feynman, Julian Schwinger and Freeman Dyson had been working on ways to calculate what was happening when electrons and photons interacted. Prior to their work some of the variables to be considered in such calculations had to be treated as effectively infinite, making the whole thing a very messy business.[14] Feynman, Schwinger and Dyson found a way to renormalize the calculations, i.e. turn them back into variables you could do meaningful mathematics with. Kenneth Wilson of Cornell University found a way to do the same for phase transitions. In truth there was no other way to tackle the problem.

Any attempt to program a computer to calculate the interactions and correlations between the particles in a hot ferromagnet is doomed to failure. Not only would it have to calculate the effect of interactions between one particle and every other, it would also have to take into account all the possible sizes of clumps at scales ranging from millionths to tens of centimetres.

An analogy may make it obvious why this approach will not work. If the system in question was not a magnet but the global weather system then any model that attempted to show what could happen would have to take in an impossible amount of detail. It would have to include large movements of air such as the jet stream, the trade winds, the Sirocco and Mistral, as well as the exhalations of everything that breathes, the draughts caused by slamming a door and the eddies caused by butterflies flapping their wings. It just can't be done.

In a far smaller system such as a magnet it might be possible to derive an answer for the first particle being considered if you had a huge amount of computer power at your disposal but that is about as far as you would get. Whatever answer was derived – which would be the sum of all possible interactions over all scales – would then become the initial variable for the second calculation, and so on, and so on. As two experts on critical phenomena have wryly put it this is a task that would resist 'even the most energetic investigation.'[15]

Wilson realized that at the critical point many of the details of the system and its parameters become irrelevant. The properties of the constituent parts are redundant. In fact there are only three numbers that matter: the number of dimensions of the system, the order parameter and the correlation length. These sound very complicated but they are fancy names for some straight-forward ideas.

Dimensions are easy to explain. There are one, two or three of them. A flat bar magnet has, effectively, two dimensions. Order, in the form of clouds of magnetism of one polarity or another, can spread along its length or across its width. A gas has three dimensions because when droplets form they can spread in any direction. Given that often the systems we are talking about are real-world systems, it is not always easy to assign an integer to describe their dimensions. Real-world systems such as mountains sit somewhere between the neat Euclidean ideals, as we shall see below.

The second number, or order parameter, called Q, captures the degree of orderliness that sets in when a system is cooled below a critical temperature. At room temperature a ferromagnet will have a Q value of almost 1 because all the grains within it will be highly ordered. The poles of the tiny grains making it up are aligned and the whole bar is strongly magnetic. When the magnet is heated, Q will diminish as the grains start flipping and this order begins to be upset.

The final number is the correlation length, denoted by ξ, and all this attempts to do is measure the size of the largest orderly structure within a system. In our magnet held at the critical point these structures are the clouds of magnetism formed by particles with the same orientation. At the critical point they can be surprisingly large. In fact they can reach across the entire system.

The order parameter and the correlation length capture the essence of a system at the critical point. They show that you must forget about measuring the orientations of the microscopic grains. Instead you should look at the size of the clouds and how they are scattered. This represents the amount of order/disorder present and the lengths over which correlations (influence) are felt. The structure, not the properties of the individual parts, is what matters. Wilson showed that at the critical point the same order parameter and correlation lengths were at work.

It might help to explain this by thinking of the magnet divided into square boxes.[16] They can be of any size but it pays to start small. Wilson was saying that if you examined an area of the magnet bounded by the tiniest box, measured the size of the largest structure and the distance over which correlations were felt, then doubled the size of the box and did it again, and again, you would get the same result. The same order parameter and correlation lengths were at work at all these scales. In the smallest box and the biggest the same distribution of structures was apparent and correlations extended over the same distance. There is a 'profound configurational similarity.'[17]

Even though you are zooming in and out, the same degree of structure or organization is apparent. In fact if you didn't know which size of box you were looking at, you would be pushed to tell the different regions apart. To anyone confronted with the plethora of phenomena that the world can throw at them this seems bizarre. Surely such things as hot water and hot magnets are profoundly different? Not according to a growing body of evidence they aren't. Examine something closely enough and you might find that it has a dynamic identical to many others. Vastly different systems can, in some circumstances and with some caveats, exhibit exactly the same degree of organization.

When systems were classified by the number of dimensions they possessed, which defined how order can spread through them, it became obvious that there were only a few categories or classes that separated them. In planar systems, such as a bar magnet, the order can really only spread in two dimensions – along the length of the bar or across it. But this was as far as the differences went. The behaviour of all the systems within one class or category was mathematically identical – hence the relevance of the term universal. Any theory that could be used to explain what was happening to one system in one class could apply with the same force to all the others in that class. The beautiful world that Kadanoff had glimpsed was not so small after all.

Kadanoff thought and Wilson proved that at critical points, which are exhibited by a huge number of systems, the same parameters show up time after time after time.

Not similar.

Not close.

Not almost.

The same.

Mathematically identical in fact.

Considered in this way boiling water and hot magnets are indistinguishable. Sand piles with avalanches of grains running down their sides are the same as other three-dimensional systems when they are at their critical point. Given the intricacy of the dynamics at work at these moments, it is not hard to see why Wilson's mathematical method to describe this won him a Nobel Prize in 1982.[18]

Now that the history is clear we can create a short list of the kinds of systems we are looking for and the properties they should possess to be considered as places that Universality might crop up.

1. The systems must be open and energy or information must be able to flow across their boundaries.
2. They must be made up of thousands, if not millions, of interacting units. Only when there are several scales available to work across does Universality become easy to spot.
3. The units should have the same sort of properties as each other. Between systems these units can be very different. Water droplets in a cloud have little in common with Wall Street stock traders, but both systems are good candidates for places where Universality is at work.
4. There must be a constant source of energy or information flowing into and out of the system driving it to a state of constantly shifting change. This energy flow 'tunes' the development of the system and gives it something to react to. In some cases this flow may be the output from another system displaying Universality.

Given these characteristics it is easy to see why the field became known as Universality. Once you know where to look and what to look for you can spot this stuff everywhere. The world is riddled with it.

With Universality physics has begun to tackle the real world. Too often physics is too aloof and unable to tackle the problems that matter. The things we are profoundly interested in – stock market movements, cycles of history, earthquakes, whatever – remain closed off. However, with Universality, the desert of our ignorance begins to bloom, and with it our understanding of the world we live in and ourselves. This is the physics of edges, interfaces, organisms and humans. Where all the interesting stuff goes on. Using Universality the universe opens right up and lets us grapple with its heart for the first time.

Boom

The important idea to take on board is that at a critical point the same degree of order is present at all scales, no matter how close or far away you get.[19] In the early 1970s, when Wilson was publishing the papers laying out the details of renormalisation group theory,[20] the systems showing this similarity across scales were said to possess a critical exponent. This expresses the degree of order apparent at any and all of the scales. In the late 1970s, thanks to the efforts of an IBM research scientist called Benoit Mandelbrot, a new name started to be used for systems that appeared to be the same at all scales. Such 'self-similar' systems started to be called *fractals*.

These shapes that go on forever started as purely mathematical constructs. When first described at the turn of the twentieth century mathematicians used to the clean hard lines of Euclidean and Newtonian geometries recoiled in horror from shapes that seemed to exist between dimensions. At first they were called 'pathological' and a 'gallery of monsters'.[21] Now we know them to be nothing more than the reflection of the real world.

Mandelbrot is a scientific maverick, an intellectual nomad who has pursued his own interests throughout his career only to find that everything he tackled was a different aspect of the same problem. It took thirty years for the realization to break in on Mandelbrot that he was seeing the same dynamics over and over again. He writes: '. . . most of my works turn out to have been the birth pangs of a new scientific discipline.'[22]

This thirty-year study began in 1951 when Mandelbrot was looking for some light reading before taking a journey on the Paris subway. He retrieved a screwed-up book review from a mathematician's wastepaper bin which mentioned Zipf's law. This law was named after Harvard University professor George Kingsley Zipf who first advanced it in 1949 in a book called *Human Behaviour and the Principle of Least Effort*. In this book Zipf tried to explain some curious regularities marking human-created systems such as the size of cities and the frequency of words in literature. Zipf noticed that there was a relationship between the size of a city and the number of other cities of the same size in 1920. The book also dealt with the shape of sexual organs and justified the *anschluss* of Austria by Germany because it improved the fit of a mathematical formula.

As he expected large cities were rare and small cities common. What he did not expect was that when he drew a logarithmic plot of the populations of cities against the number of cities with that population he would get a straight line. Rather than there being a random distribution of city sizes, it looked like someone had specified how many of each size city there should be. The numbers were that predictable. There were a couple of cities with populations over 8 million, ten with more than 1 million inhabitants and a hundred with more than 200,000 dwellers.[23]

The results were utterly different to the pattern produced by many other human systems. Typically if you measure any aspect of a system that humans create or are part of you get a spread of results called a bell curve. It is also called a 'normal' distribution, which is a bit of a give-away. If you measure the height of a class of schoolchildren you will find a few very tall kids and a few very short ones but most will be about average height. Zipf expected the same for cities but instead there seemed to be a relationship between the size of the city and the number that the world could support.

He found the same strange relationship when he looked at how often words were used in works of literature. He counted how often words were used in *Ulysses* by James Joyce and a collection of American newspapers. An equally strange relationship emerged. The word *the* emerges as the most frequently used word and has a frequency of around 9 per cent. The word *I*, ranked tenth, has a frequency of 1 per cent, and the word of rank 100, *say*, has a frequency a tenth of that. The relationship held true no matter which text Zipf examined.[24]

Mandelbrot began to analyse Zipf's law but soon he moved on to fluctuations in cotton prices, stock market movements, the staccato clusters of interference on telephone wires, and then on to coastlines and clouds. Gradually it dawned on Mandelbrot that these were not separate phenomena but shades of the same dynamic that coloured the entire world. As such they could be explained with a single idea – the fractal. He coined the term from the Latin adjective *fractus*. This derives from the corresponding Latin verb *frangere* that means 'to break' or to create fragments. A fractal, therefore, is something that is shattered yet retains a certain symmetry.[25]

The Ragged and the Real

To get a better idea of what this means think about how you would measure the length of the British coastline.[26] At first glance this seems a very straightforward task. You could do it quickly by using an atlas and some string. You could put the end of a ball of household string on Blackpool and then roll it out along the coast, making sure it follows its coves and promontories as closely as the flexibility of the string allows. If the string is quite thick there could be a few coves in Scotland that you can't bend it around exactly but you could get close enough for an estimate. When you get back to Blackpool you can measure the length of the string it took on scale of the map and get a crude idea of how long the coastline is.

However, this is not the only measure that you could take. If you use a larger map and fine fishing line that can be fitted around the features of the coast more easily and perhaps pins to mark the inlets and promontories you'll get a better answer. You would get a different, much more accurate, answer if you paced out the whole coast for yourself and counted your steps. More accurate because your units are smaller but not necessarily more correct. Which answer you will accept depends on why you are doing the measuring and what you need the result for.

A different method of measurement produces a different answer but none of them is wrong. All of them are right. In a very real sense every measure of the coastline of Britain, whether done with string or strides, is correct. As Mandelbrot says: '. . . coastline length turns out to be an elusive notion that slips between the fingers of one who wants to grasp it.'[27] There is no one definitive scale.[28] It all depends.

Just because you use different methods to measure the coastline does not mean that there is no affinity between the scales you choose. At all scales a degree of order is preserved. If you compare images of the British coastline, or any coastline for that matter, taken from different heights you would see the same degree of roughness in all of them. No matter how far away or close you are the coastline looks equally jagged.

Mandelbrot realized that a myriad of real-world phenomena can be described as fractals. Phenomena that, when examined at a variety of scales, display the same degree of order. They are self-similar. Self-similarity means that there is not one scale that can be used to get a definitive answer about the length, size or distribution of these phenomena. They look the same at all scales. A magnified part looks like the larger section it is taken from.

When fractals first appeared in formal mathematics they were viewed with horror because they seemed to break all the rules about dimensions. Geometry is all about structures that exist in one, two, three or more dimensions. It works with points, lines, planes and solids, but the first fractals – which gained the glorious names of Sierpiński carpets, Koch snowflakes, Menger sponges and Cantor dusts – seemed to be between dimensions. At first mathematicians had no idea what to do with them.

Mandelbrot realized that much that was interesting in the world existed between dimensions. A coastline is neither a plane nor a solid structure, it exists somewhere between the two. He came up with the word fractal because he wanted a term that could do justice to the ragged shapes and structures he was studying. Nothing in formal geometry, which dealt with constructs of exactly one, two or three dimensions, could be used to describe a something as messy as a coastline or a cloud. He needed a word that could do justice to structures that had to be considered at all scales not just one.

We have been here before. We have seen the same situation in a ferromagnet held at 770°C. At the critical point all systems exhibit the same degree of order at all scales. There is no one scale that matters more than the others – they all are equally important. Ferromagnets at 770°C are self-similar. So are fractals. They are one and the same thing. Wherever you see a fractal you are seeing a system at its critical point.

As well as being the man who gave fractals their name and showed how they might come about, Mandelbrot has been one of the keenest cataloguers of the places that fractals turn up. His 1977 book *The Fractal Geometry of Nature*[29] is a cornucopia of examples of fractals in the real world. It is a daunting book to read, but there is no better place to gain the insight that the world is riddled with fractals.

Fractals or self-similar structures can be found everywhere in the real world: the curling shell of the nautilus, the overlapping plates of a pineapple skin, the coastline of Britain, mountain ranges, and even the scattering of galaxies, clusters and superclusters across the universe. There are fractals at work in music, the formation of fern leaves, rainfall patterns, fiords, DNA and the branching blood vessels in your body. Some have even suggested that turbulence is fractal and that lungs are simply turbulent patterns cast, not in stone, but in cells.[30]

Moreover, it does not stop there. Fractals are distinguished by the fact that they have no characteristic length scale. In addition, there are also fractals that have no characteristic time scale, i.e. the pattern of events looks self-similar on all time scales,[31] whether you are studying a second, a minute or an hour. These fractals go by the slightly different name of $1/f$ (pronounced one-over-f) noise. They too are popping up everywhere and have been found in a huge variety of diverse systems, such as the flow of the Nile river, the pulses of light from quasars, the bunching of traffic on motorways, heartbeat pulse patterns, earthquake tremors, and even the way people surf the Internet.[32]

The defining mark of both fractals and systems that display 1/f noise is a power law that expresses the relationship between events or structures of different magnitudes. If we go back to our ferromagnet hovering at the critical point and work out the size of magnetic clouds with the same orientation, we would find a power law relationship between them. Small clouds are common, medium-sized ones less so, and large clouds are rare. There is, however, a characteristic ratio at work that dictates the number of large versus small clouds. A power law is a mathematician's way of saying that one measure N (e.g. the number of clouds) can be expressed as a power of another quantity s (e.g. size of the clouds). Drawing a logarithmic plot of the number of clouds against the size of the clouds would reveal a straight line. A power law is a purely statistical relationship. It will not tell you what will happen next, i.e. it cannot be used to predict whether a big quake will follow a small one or when a large stock market crash is on the cards. So if you are reading this book with a view to becoming as wealthy as Bill Gates you will have to think again. All it can do is tell you the distribution of events of all sizes. Nevertheless, as later chapters will show, even that can be useful.

Bak to the Beach

Power laws can even be found in the size of the avalanches sweeping down the side of the sand pile we created on the beach. If you had been able to measure the magnitude of each avalanche tripped by the grains of sand you were adding, you would find a power law at work. The pattern of avalanches looks the same across all time scales.

The sand pile model was first introduced by Per Bak, Chao Tang and Kurt Wiesenfeld.[33] The trio proposed that this simple model contained a profound lesson for anyone interested in the dynamics of self-similar systems.

The work of Kadanoff, Wilson, and the other early pioneers of renormalization group theory helped dissect the dynamics of a system at a critical point but it had little to say about how such systems arose in the real world. It had even less to say about how such systems maintained themselves outside the lab and without a convenient way of adding energy to them.

Bak, Tang and Wiesenfeld thought they had found the way. Their idea was beguiling in its simplicity. The trio claimed that systems made up of thousands, if not millions, of interacting units managed to maintain themselves along the critical boundary. This poised state comes about purely through the interactions between the elements of the system, be they sand grains, stock brokers or snowflakes. At the critical point it does not matter. The interaction drives the system to self-organize. Change when it comes, comes swiftly. Once it has passed, the system settles into its old critical state. They gave the theory the name of self-organized criticality.

By going back to the sand pile we can see how this happens. Adding grains of sand drives the pile towards its critical state. When the weight of grains we are adding becomes too much for those below an avalanche occurs.

Once the change has swept across the system the sand settles down again into its critical state. By adding more sand we drive it again until another avalanche occurs. Between avalanches the sand is hovering around its critical point. It has become self-organized around its critical point, poised on the cusp of change and ready to react.

Bak, Tang and Wiesenfeld argued that this simple model could be used to explain the origins of all the systems where $1/f$ noise and fractals were apparent. At the critical point the properties of the individual elements cease to matter and the interactions take over. Order emerges and the world rolls forth. Life begins.

Chapter Three

The Midwife of Creation

If the forms selection chooses among are generated by the laws of complexity, then selection has always had a handmaiden. It is not, after all, the sole source of order, and organisms are not just tinkered together contraptions, but expressions of deeper natural laws. If all this is true, what a revision of the Darwinian worldview will lie before us. Not we the accidental, but we the expected.

At Home in the Universe, Stuart Kauffman

If I were called in
To construct a religion
I should make use of water.

'Water', Philip Larkin

Everywhere

Long before life had the limbs to consider rolling or strolling forth it was swimming, wriggling and evolving in the early oceans of the earth. Evidence is mounting that it was only thanks to the dynamic double act of water and the principle of Universality that self-sustaining organisms, life, got going in the first place. The salinity of the blood of whales, humans, mice and most fish, as well as many other organisms is the same: 0.16 molar (around 1 per cent by weight in water). This perhaps reflects our common origins – the ancient oceans that gave birth to us all.

Water is the universal solvent that made life possible, thanks largely to the anomalous properties it displays in its liquid form. You should be glad that water is weird in this way because if it weren't you probably wouldn't be reading this now. In fact you wouldn't *be* at all and the chemicals so carefully arranged into your elegant form would be swilling around the oceans with the rest of us. Barring the odd excursion into the clouds we'd all be part of a thin salty broth with nothing to do with our time but wash up on the odd beach. Instead here we are with our books, beverages and comfy chairs, and it remains an unanswered question as to how we got from broth to books, from beach to beer and here from there.

Almost all of the curious properties of water – such as the fact that it has a higher melting point, boiling point and surface tension of comparable substances such as ammonia – are due to its chemical structure. In particular these properties can be traced to the way that water holds itself together via the complex interactions of hydrogen bonds. It used to be thought that these hydrogen bonds were simply the result of electrical attractions between a positively charged hydrogen atom and negatively charged oxygen atoms. The hydrogen atom readily gives up its single electron in water and an oxygen atom in a neighbouring molecule is happy to take them up. The result is a weak bond.

In the water molecule two hydrogen atoms connect by a single chemical bond to an oxygen atom. The elements share electrons in covalent bonds forming a water molecule. The molecules formed are bent and highly charged with one part negatively charged and the other positively charged. It looks like a paper dart with the oxygen at the tip and the hydrogen molecules swept back and away at the tips of the wings.

The polarity helps water molecules arrange themselves into a loose pattern with negative ends of molecules attracted to positive ends of other molecules and vice versa. The result is interleaving arrowheads with the hydrogen in one molecule forming weak bonds with the oxygen in another. This short-range order explains most of the anomalous properties of water.

These hydrogen bonds are not very strong but they do account for the relatively high surface tension of water. They break and reform often as the molecules jostle around but they are strong enough to stop the molecules completely dissociating, which is why water has such a high boiling point.

Water then is a shifting seething mass of molecules constantly forming and reforming hydrogen bonds making the substance sticky and highly charged. It is at this point that Universality kicks in and helps to get things started.

The highly charged nature of water means that tiny islands of chemicals can form spontaneously. This sorting happens thanks to long skinny molecules known as lipids. These ubiquitous compounds possess water-hating (hydrophobic) tails and water-loving (hydrophilic) heads. The tails hate water because they are made of uncharged carbon and hydrogen chains. By contrast the heads are made of charged sugars and phosphates and have an attraction for the water molecules that have a charge of their own. This means that in water lipids tend to organize themselves into sheets made up of two layers of lipids poking in opposite directions. The water-loving heads of the two layers point towards the outside and the water-hating tails huddle together on the inside.

Chemicals could, and have been shown to,[1] become trapped between lipid layers forming vesicles, tiny reaction chambers protected from the depredations of the wider world outside. These could have formed the first protocells or been self-contained reaction chambers that protected the delicate loops and polymers that got life going.

At low temperatures the membranes pile up in parallel layers and above a critical temperature they separate. Between these two they are like loosely bound bags that can ripple, reform, open and close to a fractal beat. Around this temperature they do not tear and shatter but will happily part and reform to let chemical complexes pass in and out. The complex play of charges found in lipid sheets, water and ionic (reactive) compounds drives this mixing. This system has many of the properties of a system displaying Universality. This principle is at work keeping the lipid layers intact yet flexible. It might have helped them enclose chemical complexes that went on to become the first organisms.

Beach Beginnings

The early oceans and beaches are good candidates for the places that these reactions might have started. The chemicals might have washed up on sandy shores and then dried out as the tide ebbed. The repeating patterns of inundation and soaking might have been enough to give them a start. No one knows.

What is sure is that so far no one has managed to recreate the conditions under which these chemicals might have formed in a lab. If these reactions are to get going then they have to be brought into close proximity and left alone for a while. If reactive or ionic compounds were allowed to float freely around the early oceans all the other floating chemicals eager to react with them would quickly turn them into inert chunks.

Universality is also at work distributing the chemicals around the oceans. Helped by the weak network of hydrogen bonds in water, compounds gradually become spread in a self-similar pattern. Like all fractal distributions this pattern has a huge surface area. It makes sure that everything gets a really good chance to react with everything else. Universality is one of the drivers of life.

Percolation, the interweaving of different molecules in scale-free distributions, is one of the most highly studied and best-established facets of Universality. There are percolation points just as there are critical points[2] at which the properties of mixed materials, such as cement,[3] change significantly. In fact the parallel between percolation and phase transitions is mathematically demonstrable.[4] They are closely allied dynamics.

At the percolation point it is often possible to trace a winding and indirect path from one edge of a system to another. If you freeze a system at a critical point you can pick all the units sharing the same properties at that moment (say magnetic orientation) and see that they form a network that spans the system. This is one of the ways that information propagates and the system sustains its ability for swift change. In the early oceans this long-range linking might have helped widely separated chemical compounds react and form larger chains. Percolation via contact points could have been a key mechanism that got those chemical compounds to sustain themselves.

A lot of research has been done on percolation but so far little of it has been applied to the chemical soup that was the ancient oceans that brought forth life. No one doubts that it was there that chemical complexes engaged and began the climb toward self-organizing, self-sustaining and living organisms.

This climb took a long time. The earliest organisms – blue-green algae and bacteria – had the planet to themselves for 2 billion years. They lack a nucleus and have their genetic material floating around inside them. They promiscuously swap genes with other bacteria. Many of the chemical tricks that these organisms, known as prokaryotes, learned over those millennia are still in use now. Humans use molecular relations of these tricks to transport oxygen around their blood stream.[5] We are united by our chemistry.

These early chemical complexes were helped by another anomalous property of water. The liquid has a very high heat capacity, which means that it takes a lot of heat to change its temperature by a degree. This is a little like putting lagging on your pipes in winter to stop them freezing and splitting. The water acts as a blanket and lets the chemistry use the energy for itself. In effect water gives a great boost to any biological or chemical reactions taking place in it because, although these processes generate heat, the surrounding water remains at the same temperature. This effectively insulates the reactions because the water surrounding them does not get too hot to interfere with what the chemical complexes are getting up to.

This heat-holding ability ensures the oceans store a lot of energy and help keep the climate cool and less prone to fluctuations. It also means that water and ocean currents carry energy in a very efficient manner too. The Gulf Stream is around 200 kilometres wide and around 500 metres deep and has been estimated to carry energy equivalent to burning 160 billion kilograms of coal an hour. These circulating currents bring energy to many parts of the ocean that otherwise would be much colder. It may have helped kickstart those early self-sustaining chemical reactions that got life started. These currents also help smooth out temperature changes providing a stable habitat for life to occupy.

Water Works Wonders

H_2O is one of the most plentiful compounds in the galaxy. Its component elements feature in the most abundant molecules: H2 and CO. It is found in every cloud of interstellar dust and in the heads and tails of comets and down here it is arguably the most important substance on earth. It covers around 70 per cent of the planet's surface. Water is the cornerstone of life as we know it.

It is involved in every living process and when we search for life beyond earth we look first for water and then work from there. Part of the reason that Mars is so interesting is because its watery past is gouged into its surface. Floods raged on Mars at some point, and if the Martian seas persisted for any geologically significant period of time then it is reasonable to expect that life may once have emerged there too. Nothing very complicated perhaps, nothing that can sip a beer and sit on a lawn chair perhaps, but alive nonetheless.

A NASA robot mission to Europa – one of the sixteen satellites of Jupiter and one of those first spotted by Galileo in 1610 – is being planned only because astrobiologists are convinced that there is water there in abundance. And where there is water, there is a niche for life to claim as its own. The *Voyager* and *Galileo* missions have taken pictures of the surface of Europa and the images they returned have revealed a pocked and cracked surface reminiscent of our own ice caps. Some space scientists believe that beneath the surface layer of ice lie oceans that are kept warm and circulating by the core of the satellite. The ice not only keeps the heat in but also protects what is below from cosmic radiation and ultraviolet light both of which can do a lot of damage to simple cellular life forms. Europa might be the perfect incubator for extraterrestrial life.

We'll have to wait a while to find out and even when the robots land there is no guarantee that they will be able to break through the ice and plumb the depths of the satellite's oceans. There is a debate over the thickness of the ice on Europa. Some think it will be quite thin and only a few metres at most; others suspect it might be hundreds of metres deep. If it is very thick then it might be too thick for the robots to dig through but NASA scientists are already building a robot submarine, or hydrobot, that will be a prototype for those that one day may plumb Europa's oceans. As yet no date has yet been set for the mission to Europa.[6]

The hydrobot is going to be first used in the chilly waters of Lake Vostok in the Antarctic to test the technologies that would be used in a mission to Europa. Lake Vostok is a body of water almost as large as Lake Ontario and it lies under nearly 4 kilometres of ice in Antarctica. Analysis of ice core samples taken from the ice above the lake suggest that it might have been sealed under the ice for as long as a million years.

Some biologists studying the origins of life on earth think that a similar ice cap might have helped simple cellular organisms survive and evolve in the early oceans of our planet. The ice on earth may have done the same job that the ice on Europa is doing now and shelter the nascent organisms beneath.

It is certainly fortunate that ice floats. When most matter is cooled it contracts. Not so with water, it expands. That liquid water is denser than its solid form is only one of the many anomalous properties of this remarkable substance. It floats because of the way that the weak hydrogen bonds that hold water together arrange themselves as it cools. It is just as well that it does. If it did not then lakes would freeze from the bottom up. This would have had dire consequences for any living organism, which would have faced the threat of being trapped in ice or exposed to the harsh temperatures toward the surface of a lake or sea.

As it is, ice has a great insulating effect and stops the temperature of the water dropping below a certain limit. If all life sprang from replicating colonies of chemicals then this insulating effect would help the colonies stay at the right temperature to react, recombine and drive their way toward simple cellular organisms.

U Are Here

Strong evidence for what kind of chemistry was present at the birth of life may be missing but the evidence that Universality was there is easy to find. Using simple models many scientists have found persuasive evidence which suggests that Universality is built into living organisms as their, and our, defining dynamic.

One of the abiding lessons of Universality is that it can be studied using simple models. At critical points and phase transitions the properties of the parts of a system stop mattering. Instead it is the interactions and organization that matter. Since the 1970s scientists have been using simplified models of the world called *cellular automata* to tackle a huge variety of research problems. Scientists studying fluid dynamics and the onset of turbulence make great use of them.

The field of cellular automata was kicked off by John von Neumann, who in the late 1940s drew up specifications for a machine that could reproduce. Initially von Neumann was contemplating actually building a robot that would use sections of girders to build endless copies of itself but he was persuaded by mathematician and friend Stanislaw Ulam to make the thing more virtual. So instead of girders and machines von Neumann imagined a vast grid, like an enormous draughtboard, on which the reproducing machine would live. The elements of the machine would be the square boxes in the grid. Each box could be in one of several states. With the right rules von Neumann believed that he could create an artificial organism that could reproduce. Although von Neumann finished the blueprints for this artificial creature he had not established if it would be able to reproduce before cancer claimed his life in February 1957.

Von Neumann's creature was fiendishly complex. It occupied around 200,000 squares of the virtual grid and each cell could be in any one of twenty-nine states. Not surprisingly it was a while before anyone took up the idea. However, in 1971 Cambridge mathematician John

Horton Conway revisited the idea, tinkered with the rules, simplified it and created a new version that he initially played with squared paper and seashells. He called it the Game of Life but it only really took off when it was turned into software.

Like von Neumann's original conception the Game of Life is played out on an infinite grid. But instead of twenty-nine states the cells of the grid exist in two: alive or dead. Time advances in steps in this toy universe and the fate of every cell at each step is decided by the living or dead cells surrounding it. If three of the eight cells abutting an empty cell are filled then in the next time step that dead cell comes to life. If a cell has two neighbours it stays alive at the next time step. If it has less than two or more than three neighbours it dies. These simple rules made it a cinch to turn into a computer program and thousands of people have whiled away hours playing with it.

Despite the simplicity of its rules the Game of Life is utterly absorbing to watch and tinker with. You don't really play it, you observe and act like a beneficient, or destructive, deity as you choose. Whatever role you take on, it is infectious. Sparsely populated screens can burst into life with all manner of block-like beings that blink in and out of existence. Some, called gliders, even stroll diagonally across the grid-world like drunken bishops.

But the Game of Life is not the only game in town. Since Conway came up with the idea other researchers have come up with versions, called one-dimensional cellular automata (1-D CA), that have simpler rules and proceed line by line rather than up and down a grid.

British wunderkind Stephen Wolfram has been one of the pioneering cosmologists of these 1-D CA universes. He has spent hours classifying the different types of patterns produced by slight changes in the rules.[7] Different rules mean taking into account larger or small neighbourhoods of squares when working out the fate of a particular cell. Wolfram chose these simpler CAs because they make it possible to grasp with a glance just what kind of pattern is emerging. They also have fewer rules (256 in fact), which makes the mathematics easier. He has found that the patterns that these simple CAs, and by implication all CAs, produce fall into one of four categories. Class I patterns go nowhere: they either disappear quickly or settle into a fixed state. Class II automata patterns grow to a fixed size and then repeat forever. Class III yield chaotic patterns that look similar but never repeat. Class IV patterns are the really interesting ones. These produce complex patterns that grow and contract regularly.[8]

Wolfram's work has been taken on and extended by Chris Langton, *the* pioneer of the field of artificial life (ALife) and now a full time researcher at the Santa Fe Institute, one of the main centres for research into complexity. Langton who, ironically for one so tied up with definitions of the living was a morgue attendant before he became a scientist, coined the term *artificial life*, did some of its first and most important work in the field, and organized the first conference on this subject. He's the man who breathed life into artificial life research.

As its name suggests artificial life is all about creating synthetic organisms. Unlike artificial intelligence research that begins with the brain and works down, artificial life begins with the boots and pushes up. Typically researchers try to get something straightforward right first, such as making a robot walk, and then build on that and see what kind of intelligence emerges when you put a few competences together. Many others use cellular automata as toy universes into which they unleash a variety of software creatures to relive and retrace evolution.[9]

Half-Empty, Half-Full

Langton's survey on CAs was more exhaustive than the one Wolfram carried out. He classified and tinkered with thousands rather than just hundreds of 2-D CAs – variations on the original toy universe that Conway drew up in 1971. He found that there was a fine line between the patterns that never changed or dwindled into nothing and those that bloomed into unpredictable chaos. Between disorder and boredom was a 'sweet spot' where the most complex patterns emerged. In this region patterns propagated, extended over long distances before being superseded. Patterns bloomed, died and popped up again. Fractal patterns, a phase transition, a critical point. Universality.

Rather grandly Langton christened this point 'the edge of chaos', the cusp between death and disaster. He could just as easily have called it 'the edge of order' but that phrase does not have the same thrilling ring to it. In some respects this is the wrong label for it because in mathematics 'chaotic' and 'chaos' have very precise meanings. Chaotic systems have readily measurable characteristics and features that can be used to pick them out. Universality is the opposite of chaos. The ubiquity of order, not chaos, is the key measure of systems at critical points.

Leaving these semantics aside it is enough to say that Langton found a small region in which the dynamics of organization and information dominated. His work 'suggests the possibility that the information dynamics which gave rise to life came into existence when global or local conditions brought some medium – perhaps H_2O, perhaps some other material – through a critical phase transition'.[10]

Langton speculates that this is the niche that all living organisms occupy. At the critical junctures he has seen complex structures emerge and persist. He goes as far as to speculate that the same thing happens in living organisms which, consciously or not, inhabit the same kind of critical point. His conjecture is backed up by evidence that only at this point does the amount of information passing through the system vastly increase. Persistence, staying alive, demands the ability to process information about your surroundings and use it. Again Langton believes that this makes the critical region a good candidate for the place that life got going.[11]

Langton suggests that at this point, the edge of chaos for want of a better term, the ability of a system to use and process information improves dramatically. Information in this sense means the ability of the different cells to influence each other and how well this is passed on or correlated. The ferromagnet at the critical point is permanently uncertain. The clouds of magnetism that form and flutter along it and within it are formed when the microscopic magnets are 'persuaded' to reverse their polarity by neighbouring units or thermal noise. This choosing – the flipping polarities in the magnet and 'alive' or 'dead' cells in the cellular automata – is what Langton means by a dynamics of information coming to dominate.

In essence the units make a decision about what they want to do based on the past behaviour of the other elements in the system. Order is preserved, propagated and acted upon. At this point there exists a tense struggle between the storage of information and its transmission. The cells in the CA struggle to stay alive (storage) but could easily be killed off by changes anywhere in the CA universe (transmission). For the cells in a CA it is a life and death struggle.

There is also a more formal sense in which Langton means that information dynamics have come to dominate in living organisms. Information theory was developed by Claude Shannon in the opening decades of the twentieth century. He formalized (worked out the mathematics for) the ways that information can get corrupted as it is transmitted across noisy channels. Noise, or interference, is unavoidable and ubiquitous. Instead of trying to design systems that never fall prey to it, communications engineers instead find ways of correcting the errors in messages that noise creates.[12]

The old playground joke about an order given during the First World War illustrates the dangers of mistakes in copying. The original, and surely apocryphal, order was: 'Send reinforcements we're going to advance.' As it was passed by word of mouth from down the chain of command it gradually became corrupted, and when finally delivered the message had become: 'Send three and fourpence we're going to a dance.'

Shannon was interested in improving the ways that data could be transmitted down wires and over the air waves by compensating for or removing interference or noise on the line, but his theories can be applied anywhere that information is being passed on. In this context information is an abstract term for anything and everything you want. It could be light pulses down a fibre optic cable, radio signals across the Atlantic, or sequences of bases on DNA strands being rebuilt during cell division or swapped during reproduction. It doesn't matter. At the critical point there is a fierce tension between the corruption and preservation of the message and the amount of information transmitted is at its maximum.

It should be stressed that this poised state is a long way from being balanced. Universality is characterized by quick changes and catastrophes. A diamond or a table is balanced; systems showing Universality are not. Quite the opposite. Tiny perturbations can have enormous consequences and change can sweep through a system in a moment. There is no equilibrium here.

At the critical point order and structure are all important. Physical properties cease to matter. It is this Langton suspects that gave rise to living organisms and, once up and running, flying or swimming, is the dynamic they use to keep going. He, and many others in the ALife movement, think that this is more than just an analogy. They see it as the defining organizational principle of all living things. They believe that organisms are riddled with critical points and because of this information, genetic or structural, is the key measure of life. Changes rippling through the system, which represent information, help the organism keep itself going. They are the motor of evolution and reveal how it is adapting to its environment. What is important about life is what it does, the process, not what it is made of.

Langton acknowledges that much of this is speculation on his part. The links between his edge of chaos idea and living organisms have not been firmly established. Some of this work suggests he is on to something. Before now his discovery and exploration of the 'edge of chaos' idea has been of mainly phenomenological interest. There has been little ground for believing that what Langton found in computational systems (the Game of Life in this case) also occurs in living systems.[13] But the idea has been of huge importance to the nascent field of ALife. The appeal of the idea has given ALife researchers an idea of what they should be looking for and what they should be trying to create.

While there is no doubt that critical regions do give rise to the same sort of dynamics that are found at the edge of chaos, the question of just what relevance these have to each other has yet to be settled. The dynamics Langton found at work in CAs may just be a pale imitation of what exists in many real-world systems. The link between life and ALife remains to be conclusively proved. This is partly because Langton did a lot of his research on the edge of chaos idea before the explosion of interest in phase transitions and critical phenomena that took place in the late 1980s. However, some recent work on populations of artificial organisms has taken his work and extended it further. The conclusions are provocative.

In the Garden of Avida

The test ground for Langton's ideas about the dynamics of life is yet another CA. This one, however, is slightly different to the one dreamt up by John Horton Conway. This CA is the creation of Christoph Adami, a physicist from the California Institute of Technology in Pasadena, and his colleagues Titus Brown, Charles Ofria and Richard Lenski from the Michigan State University. They called their creation Avida and again used a limitless grid as the stage for the action.

However, to build their toy universe, Adami and his colleagues have drawn directly on the work of another ALife pioneer, Tom Ray. Softly-spoken Harvard-educated Ray is an ecologist and biologist by training and inclination who took up ALife because he was frustrated with the work he was doing in biology. Now he carries out his research at the Advanced Telecommunications Research labs in Kyoto, Japan.

Before he was an ALife pioneer Ray was a field worker for the eminent myrmecologist E. O. Wilson, who employed him on ant-collecting trips to the Costa Rican jungle. Ray developed a real skill in digging out the heart of leaf-cutter ant nests that are made up of the bodies of the ants themselves. While in the jungle Ray noticed the parade of insects, birds and other creatures that were following the foraging ants as they searched for food. This piqued his interest in ecologies and evolution and how such symbiotic relationships could develop. His time in Costa Rica also spurred some work he did on giant vines that seek out shadows instead of the light.[14]

Ray grew frustrated with the straightforward naturalism of this type of biology. Getting to know everything about what was happening now told him very little about how such things had developed. The inability of ecology to explain how symbiosis emerged and its failure to show how evolution was playing out led him out of the forest, back to the lab, and to a world that he created for himself in a laptop computer. He called this world *Tierra* (Spanish for Earth) and seeded his creation with chunks of computer code around 80 bytes long. The creatures were made of half-formed computer instructions and among these was the ability to copy, or reproduce. Soon after the program was set off and a few energetic generations the computer code creatures filled up the chunk of computer memory Ray had set aside for their little world.

Ray forced evolution to work quickly in this programmable primordial soup by occasionally scrambling, or mutating, the instructions as the 80-byte cyber-creatures reproduced. He likens the process to the shuffling of genes that goes on when real organisms reproduce. Ray himself has seen similarities between the programs in Tierra and lengths of nucleic acid. He used a special program called the 'reaper' to clean up the worst-performing mutants, and occasionally the older ones too, to open up space for the new-born and newly mutated programs to expand into. There is a random chance that it will operate every time a program reproduces. It strikes automatically when a particular program has gone through a set number of generations. Without the 'reaper' Tierra would have ground to a halt after only a few generations; with it everything that was born was given a chance to succeed.

When he first set Tierra going Ray was not sure what would happen. He thought that he might have to tinker with it for a while to produce the evolutionary dynamics he was interested in. However, instead of grinding to a halt as it filled up with a series of tiny computer programs that did nothing, Ray found that, right from the start, raw evolution could be seen at work in Tierra. The gentle pressure of evolution turned the half-formed instructions into working programs. Populations of fitter organisms grew, multiplied rapidly and then dwindled back to a more stable level. Parasites sprang up that hijacked the codes of more complete creatures for their own reproductive ends. The populations of the toy creatures fluttered, bloomed and died away just like those seen in the real world and in the fossil record.

The success of Tierra – which is now being extended to run across the Internet so the creatures have new habitats to explore and fresh evolutionary pressures to shape them – has spawned an explosion of copycat programs. Avida, created by Adami and his colleagues, is just one. In Avida though the researchers have made some significant changes that make it, in their minds, more representative of the real world.

For a start they have restricted how and where the reaper program can act. In Tierra it is an omnipotent entity that can remove programs from anywhere, creating space for new organisms to move into anywhere on the CA grid. To Adami and his colleagues this seemed too divorced from the real world. The success of an organism is reflected close to home not anywhere across the globe. As a result when the reaper acts in Avida it can only create space in the eight squares surrounding a reproducing cell.

The biggest change was made to the organisms themselves. In Tierra an 80-byte creature took up 80 bytes of memory, but in Avida creatures were discrete. They fitted in one cell on the limitless grid and their genome was stacked vertically like a huge tottering hat. If the grid were three-dimensional these instructions would poke out of the screen. Again this genome takes the form of computer instructions. What takes place when this toy universe is left to run is a curious combination of the blinking liveliness seen in the Game of Life and the pulsing population surges of Tierra. Mutation and natural selection bring about the evolution of organisms better adapted to this mock world.

It was in Avida that Adami and his colleagues tested the idea that the dynamics of information storage and transmission are the drivers of evolution. In the context of evolution and DNA they took the storing of information to mean a longer genome. Passing this information on to new generations is the same as transmission. By varying the mutation rates they could see the point at which these opposed forces became locked. Too much mutation and any important messages for the next generation would be shattered and scrambled before it could be passed on. Garbled information is no good to anyone; but if there are not enough changes organisms would stagnate and never develop. To be effective evolution has to be able to play with the genetic make-up of creatures. Adami and his co-workers wondered at what point the computer creatures in their electronic petri dish would maximize both. If Langton was right they should see the best adaptation when critical phase-transition-like dynamics were present. That's exactly what they did see.

As the mutation rates were turned up they saw the population of organisms gradually improve in their ability to retain and pass on information. Beneficial changes were preserved and passed on to new generations. Soon, however, the rate of adaptation peaked and then fell away swiftly as the mutations and errors in coding multiplied, and organisms found it impossible to hold themselves together[15] as chaos quickly ripped them apart. Between these two extremes they saw at work everything that Langton claimed was happening in the real world. At the critical point or on 'the edge of chaos' organisms stay alive by preserving what is best about themselves but at the same time remain open to the possibility of change. It was evolution in action.

A Critical Flaw

Adami and his colleagues went further than just saying that their work with Avida lent support to Langton's ideas about phase transitions. They claimed that it proved there was a direct link between these CA models and the real world. The Universality seen at work in the world and in the CA were one and the same thing.

They bolstered their claim by drawing on the work of Danish physicist Per Bak, who in 1987 wrote a paper that almost single-handedly rekindled interest in critical phenomena. The avalanche of research and debate that it caused is rolling still. The paper[16] was published in *Physical Review Letters* and was co-authored with Kurt Wiesenfeld and Chao Tang, physicists who worked with Bak at the Brookhaven National Laboratory on Long Island, around sixty miles east of New York. Many of the 3000 scientists and technicians working there do particle physics, but Bak and his colleagues were part of a small research group that was investigating condensed matter. This is just a fancy name for stuff that isn't liquid. Physicists who study condensed matter try and work out what happens when you heat, cool or otherwise experiment with substances such as superconductors.

The Bak, Wiesenfeld and Tang paper made a very bold claim. It suggested a mechanism for $1/f$ noise, the self-similar pattern of events found in a huge variety of real-world systems. This $1/f$ or 'flicker noise' is found in such diverse systems as the fluctuations of the water level of the Nile, the electromagnetic radiation emitted by quasars and the bunching of traffic on busy motorways. Many physicists had studied the origins of $1/f$ noise before the Bak, Wiesenfeld and Tang paper but they had dealt with each example of it on an individual basis. No one had ever tried to create a single theory that could account for them all. No one had ever thought that there might be one cause behind them all.

This is what this trio of physicists were proposing. They saw no need for separate theories and instead said one idea, one dynamic could explain them all. They gave the idea the name of *self-organizing criticality* (SOC). This is a formidable name for a very simple idea. The contention was that any system made up of thousands if not millions of interacting elements, be they cars on a highway or atoms in a star, will organize itself into a poised or 'critical' state – critical in the sense of second-order phase transitions seen in ferromagnets held at 770°C. This arises solely through the jostling interactions of the individual elements. Movement and change is the essence of this 'critical' state. It is a restless shifting state that is far out of balance. Tiny disturbances can cause huge changes that extend their influence throughout the system, structures persist by virtue of this interaction. Anyone plotting the magnitude of these changes will see a power law or self-similar distribution.

Self-organized criticality is really just another name for the second-order phase transitions that this book is all about. Universality with a different hat on. However, the majority of physicists investigating critical phenomena are still debating how some critical systems reach that phase transition. In classical models, such as the ferromagnet, it is obvious what is driving the system towards its critical point: the experimenter. By keeping the temperature steady the magnet sits in the knife-edge between order and chaos. However, in real-world systems, such as earthquakes, coastlines and trees, there are no experimenters and no obvious lab apparatus they can point their finger at. They have few clues about how the fractal pattern of events is generated. Bak and his colleagues claimed to have found the engine of this dynamic. It was simply the interactions of the individual elements that brought this critical state about.[17]

The simplicity and appeal of the idea is evident in the system that Bak adopted as his leitmotif. To grasp just how the self-organization works imagine yourself playing with the pile of sand mentioned in Chapter 2. In critical systems change happens suddenly – there is no gradual evolution towards a single state. Because he chose a sand pile as the canonical model Bak called these sweeping changes avalanches, but they occur in any system displaying Universality. When the pile starts to be plagued by avalanches Bak claims that it has reached a state of self-organized criticality: self-organized because the grains of sand do it for themselves; critical because at this point avalanches of all sizes will be seen. Anyone measuring the magnitude of these cascades will see that they obey a power law, just like those real-world systems. Out of a few handfuls of sand has emerged a system that is capable of generating structures, fractal distributions, of immense complexity.

A Sweeping Claim

The power of this idea is its simplicity. Because there are so many wildly different systems that display the hallmarks of Universality any theory that is going to try and explain how they get that way has to be general. As has become clear the details of systems are unimportant; it is how they are organized that is key. The problem with all theories that try to explain what is happening in complex systems, which we can take to mean those displaying Universality, is their generality. Because they apply to such a vast range of phenomena they cannot be built around the details of any one system. One size must fit all.

Worse, they are purely statistical theories. They will never tell you what will happen next or when the next big avalanche is due. All they will do is give you an idea of the spread of events. Complexity is a theory of process not a detailed account of accidents. Bak's theory provided a general framework for thinking about and studying systems displaying Universality. It was one of the most cited papers of 1987.

Statistical theories are well known in physics and have proved their worth many times. The early work on gases was done under the framework of statistical mechanics, which does not care about details but deals only with the aggregate behaviour of the molecules or atoms in a gas. Quantum mechanics and chaos theory are also statistical theories.

Sand Pits

Bak and his colleagues did much of the initial work to establish SOC on computer. Because they were modelling critical points they knew that they did not have to construct a very complicated simulation. All they needed was something that would capture the dynamics and could be used as a general model for all such systems. The scientists suggested that $1/f$ noise and other fractal distributions of events were all brought to the critical point by self-organizing. Interactions between the individual elements gave rise to the familiar pattern of events.

Unfortunately when experiments started to be done with real sand piles problems emerged. The computerized sand pile exhibited all the qualities of a system at a critical point but real ones didn't. The power law distribution of avalanches did not appear. The 1987 paper created a flurry of experimental activity as scientists all over the world rushed to confirm its findings with real sand. Scientists seemed to see self-organizing criticality everywhere.

Leo Kadanoff, Sidney Nagel and Heinz Jaeger at the University of Chicago used a rotating drum of sand like that found in a cement mixer.[18] As the drum turns the sand slips to the bottom in a series of never-ending avalanches. Unfortunately measuring the avalanches revealed no power law. The problem was the effects of inertia that Bak and his co-workers had not included in the simulation. As the heavy grains roll they pick up momentum and get carried a long way. This means avalanches persist and swamp any signs of Universality. All Kadanoff, Nagel and Jaeger got were lots of small avalanches and an occasional huge one. There was no sign of Universality.

At IBM's Yorktown Heights research centre Glenn Held and his colleagues devised an elegant experiment to look for power law behaviours. Held built a sand pile on top of a circular plate 2 inches in diameter by slowly dropping grains onto it down a slowly rotating glass tube. A computer was set up to measure the amount of sand falling over the edges. The plate rested on top of a scale that worked out the change in the weight of the pile as the sand released in the avalanches cascaded over the sides of the plate.[19] Held found the same thing that Kadanoff, Nagel and Jaeger did: no sign of self-organization, power laws or Universality.[20]

Jens Feder, Torstein Jøssang and colleagues gave the theory the greatest chance. The scientists from the University of Oslo used rice bought from the local supermarket instead of sand to see if the lighter grains would be less prone to inertial effects. Long-grain rice seemed to work the best. Other types of rice were too ball-like and slipped over each other too quickly. They built their pile by dropping rice between two plates of glass. A camera hooked up to a computer took a snapshot every 15 seconds to see if the rice had moved. Each run of the experiment lasted around 42 hours and over a year of data was collected.[21]

Unlike the other experimenters Feder and Jøssang did see power laws, but only extending over a couple of orders of magnitude. The results were suggestive but not conclusive.[22] So were those found by Hungarian physicist Tomas Vicsek, who looked for power laws in slumping piles of mud on a miniature landscape.[23]

Bak has explained these somewhat inconclusive results by saying that laboratory experiments do not run for a sufficiently long time to produce sufficient results to give sufficiently good measures. The power law distribution implies the larger an event, the rarer it is. To see an avalanche of size 1 million you may have to wait for and record a million of size 1. That can take a long time.

Inertial effects seemed to be getting in the way too. So much so that Pierre Evesque of the École Centrale suggested putting a sand pile experiment into space to see what happened in zero gravity and tease out the effects of inertia. The pile would be in a centrifuge so that a range of gravitational fields could be studied. The European Space Agency has yet to take him up on this suggestion.

It seems to be the case that power laws only emerge in real-world examples of the sand pile experiment if extreme care is taken. Unless an experimenter pauses after each avalanche all they get is the occasional large slump. Without that tuning you get nothing of interest. The sand piles are ticking periodically rather than showing any self-organizing behaviour. The explanatory power of SOC lies in its generality and its insistence that it needs no special conditions. It is just supposed to happen. Yet, as more and more experiments are showing, it needs care and attention to bring it about when applied to real sand.

If you have a badly ventilated bathroom then every

time you have a shower you may be rerunning one of the few experiments where self-similarity has been found. Charles M. Knobler and Daniel Beysens have found that condensation forms on a mirror in a fractal pattern. The pair studied droplet formation on a mirror and measured the radii of isolated droplets. They found that the size of the droplets conformed to a power law. Moreover, although water in the form of steam was constantly being added, the same proportion of the mirror, around 56 per cent, was covered by droplets all the time. As droplets grew they started to coalesce and run off leaving room for others to form.

Henrik Jeldtoft Jensen, an expert on self-organized criticality, thinks that it is too soon to issue a decision on the theory. Ironically it appears that Bak may be mistaken about sand but he has highlighted the properties that many other systems share and the dynamics they exhibit. Universality is the first significant theory to emerge from several decades of work on complexity theory and it is thanks to Bak that it is receiving such energetic attention. He says the science of complexity is at the point physics reached at the turn of the nineteenth century. At that time a lot of basic work had been done characterizing the world but it was still open as to which scientists and theories were on to something. Now, as then, there were a lot of ideas around and the work is going on to sort out which ones work and which don't. Whether self-organizing criticality survives or not remains to be seen.

The experiments described above and many others have dented the credibility of SOC as a theory. At best they imply that the idea remains to be proved and at worst that it is non-existent. A theory that claims to be ubiquitous should be just that, but as the evidence is gathered self-organizing criticality is being revealed as far from commonplace.

This conclusion has implications for Universality as a whole. Despite the rather grand name of 'Universality' none of the scientists working in this field is pretending that what they are researching will explain every complex system. The dynamics of many complex systems, such as the weather, have been laid bare long before now and no one need look to Universality for further explanation. Chaos theory, one of the intellectual cousins of Universality, is just one example of a theory that does a great job of explaining what is happening in some discrete systems.

Viral Rights

If self-organizing criticality has shown that systems displaying Universality are not ubiquitous, they do turn up with alarming regularity. Even if Bak was wrong about the mechanism that drives sand piles to their critical point, he was right about the signature of such systems. Universality may have helped self-perpetuating chemical complexes to keep going; it may have helped the first crude cell walls form; and it may even have helped the first organisms get going.

Stuart Kauffman, one of the world's experts on complex systems, and a good chemist to boot, has done a lot of work at the Santa Fe Institute and the University of Pennsylvania on the origins of life. He is one of the few people who has tried to model what might have been happening in those early oceans. In contrast to many other researchers Kauffman thinks that the emergence of life is not a chancy thing at all. Far from it. He believes that the chemical networks and relationships that were established in those ancient oceans were almost bound to produce structures of greater complexity and, eventually, living organisms. He reasons that billions of years of evolution is not going to produce something sickly and easily upset. Instead it is likely to produce a system that can soak up a huge amount of damage and sustain itself in the face of constant and challenging change.

Kauffman's interest in chemical complexes grew out of his work on genes – a perfect example of a highly ordered system about whose origins no one has a clue. To test his ideas he created a network of 100 simulated genes. Each gene could be switched on or off, so in total the system could assume 2^{100} states. If the network were to wander through these states then, assuming it visited one state every millisecond or so, it would take more than a billion times the age of the universe to go through all the possible permutations. The network of genes we carry around with us is much more complex than this and would take correspondingly longer to check out all the permutations. Kauffman thought that once you got a network running order emerged for free. With his 100-gene model the network of genes quickly organized itself into cycles that lasted for an average of ten states.

Kauffman did the early work on his ideas using an

IBM mainframe and a 100-node network that he turned into punched cards so the computer could process the program. The nodes were switched on or off (representing information flow) depending on whether other, connected, nodes in the system were on or off. The logic of the program determined which node was on or off at each time step. At the start of each program run every node was randomly connected to two others and the rules which determined which node turned on or off at each step were randomly assigned too. To make matters more random Kauffman shuffled the cards before they were fed into the hulking machine. Usually this is a recipe for producing no results at all. Biology may be robust but computers are not. They can be upset by the smallest thing. Instead of no results, Kauffman's random shufflings produced a system that did manage to sustain itself through many cycles. Shuffling the cards every time the program was run did not seem to have any impact on its ability to find a network of nodes and keep information flowing through them.

Kauffman went on to experiment with genes that network with three or four others instead of just two. He found that as the number of interactions goes up the system goes through a phase transition in which cycles persist for an arbitrarily long time. There is evidence that real genes only interact with two or three others, so Kauffman's work may have uncovered a general property of the whole system.[24]

From genes Kauffman is now turning his attention to chemical complexes to see if small networks of them show the same dynamics. In his words he is looking for the autocatalytic sets that once set in motion drive themselves towards ever greater organization. His early work on these sets has revealed that they exist at a critical point. Put enough of them together and you get much more than you bargained for. As yet his ideas are still at the tinkering stage. His claim that this is what life does is largely anecdotal. His interest in the field stems from a conversation he had in a hot tub with another scientist friend.[25] However, there is evidence from other quarters that critical points are important in the origins of organisms.

Viruses are among the simplest of all life forms. Most are little more than a chunk of genetic material such as DNA or RNA covered with a protein sheath. By itself the virus has no way to perpetuate itself. To keep itself alive it hijacks the genetic machinery in the cells of other living creatures. Some scientists speculate that these crude organisms – little more than stretches of chemicals looking for a free ride – were the first proto-organisms. It is possible to synthesize chunks of RNA and DNA that, given the right conditions, reproduce and stay alive. It is an open question as to whether those conditions arose in the early oceans. The original primordial soup might have been a very promiscuous broth with chunks of chemicals within it cruising around looking for the length of genetic material that was their Mr Right. Polygamous parasitism was the only game in town.

There are similarities here with the *laissez-faire* reproductive technique of prokaryotes. These are the blue-green algae and bacteria that lack nuclei but mastered many of the chemical tricks that large living organisms still practise. These organisms happily swap genetic material with any and every other bacterium they meet.

RNA is key to all this because of its flexibility. By contrast DNA and proteins are one-trick ponies – they do separate jobs very well. RNA has mastered both. It can both copy information and act as a catalyst. By contrast DNA is inert and unreactive; it is unbeatable at copying information but useless when it comes to synthesizing chemical reactions. The reverse is true of proteins. Like a dating agency they are masters at getting reactions going, but they are rubbish at preserving or passing on information. RNA can do both. It can both make things happen and, with the right raw materials, make copies of itself to share the work. Never mind the chicken and the egg: RNA was there before both of them. It is the leading candidate for the original self-replicating compound that gave life its start.

It is likely that it picked up these tricks in the early oceans of the earth and it has not left them behind. There is little that goes on inside cells that RNA is not involved with in one way or another. RNA-dependent enzymes take messages (in the form of RNA) from genes. Ribosomes, cellular machines driven by RNA, translate this message and pass it on to another RNA molecule that prepares the amino acids that will turn the genes' message into action. RNA can even cut and splice other molecules, arranging them to its own satisfaction. It looks like it can do this thanks to Universality.

There is evidence that small lengths of RNA can exist at a critical point. It might be the dynamic that drove them to grow into ever larger lengths. Ricard Solé at the

University of Catalonia in Barcelona has found that RNA viruses flourish close to a perilous threshold called the error catastrophe.

The only way that RNA viruses can preserve themselves is by reproducing and adapting quickly; effectively they do this by travelling in disguise. Usually they are beaten off when the immune system of the host they are infecting recognizes them and stops them hijacking the bodies' reproductive mechanisms. To stay one step ahead the RNA viruses have to mutate swiftly and change costumes at a moment's notice. They do this by creating a cloud of mutants around a master sequence of the original virus. These mutants or 'quasispecies' preserve the nasty parts of the virus but with a few changes that might fool an immune system. To continue the costume metaphor they might don a ridiculous moustache or a fez, a disguise just sufficient to fool the rather stupid host they are trying to exploit.

Solé and his co-workers found that RNA viruses show all the signs of a system at a critical phase transition. The mutation rate of the viruses was very high, almost enough to rip them apart. Almost. The high mutation rate meant that the virus was generating enough quasispecies wearing different costumes to keep its host fooled and ensure it was preserved. Perhaps a similar mechanism was at work in the ancient oceans and helped lengths of RNA and DNA create enough mutants to survive, prosper and evolve.

The Road to DNA

It does not stop there. Living organisms have had a companion as they made their way from their murky beginnings as simple self-sustaining molecular structures to RNA and then DNA. The companion is Universality. It is not just RNA viruses that seem to exist at this edge point, poised in the critical region. Some recent work suggests that RNA and DNA are very closely related. This is perhaps not surprising if, as some scientists studying the origins of life think, ribonucleic acid (RNA) was a precursor to Deoxyribonucleic acid (DNA).

IBM scientist Richard Voss set off a lot of this work in 1992 when he carried out some analysis of the genome of an organism called cytomegalovirus – strain AD169 of it to be precise. Analysing the genome of this and 50 million nucleotide positions in more than 25,000 DNA sequences confirmed that some kind of fractal organization was at work. The suggestions as to what end are tantalizing. He found that correlations extend over larger distances in higher organisms.

Everyone knows that the instructions to build organisms are found in the strands of DNA coiled in every one of their cells. The coils are wound pretty tight – so much so that around 6 feet of DNA is squeezed into each cell.[26] Despite the huge amount of information it takes to create giraffes, geraniums and Germans, and indeed every other living organism, DNA itself is a relatively simple molecule. In fact organisms are described by an alphabet of only four letters: A, C, G and T. The letters stand for adenine, cytosine, guanine and thymine. Each one is known as a base. The bases are attached to a sugar backbone and together the two are known as a nucleotide. The nucleotides are joined together by phosphate bonds that link to the sugar groups in the next nucleotide. This mechanism binds the nucleotides together into a long polymer or chain that has the deoxyribose sugars as its backbone.

In cells DNA exists as a double helix with two sugar backbones on the outside and the bases within. The bases are picky about who they bond with. Adenine will only pair with thymine and guanine only with cytosine. This pairing preference helps DNA preserve and recreate the information in its long chains. These bases can also be divided into purines or pyrimidines. Adenine and guanine are purines and cytosine and thymine are pyrimidines. The names are used to capture structural differences in the nucleotides. Both are organic molecules that contain carbon and nitrogen atoms but differ in how these are arranged. Purines have a double-ring and pyrimidines a single-ring structure.

Replicating the information in the chain is easy with this mechanism. Since adenine only pairs with thymine and guanine only with cytosine, it is easy to rebuild a chain when it is unzipped. Just ensure that the pairing preferences are respected, and the chemistry means they have to be, and you can recreate the chains. The likes and dislikes of A, C, T and G make this a much more straightforward process than it otherwise might be.

The information in the chain is written as a series of bases. A triplet of bases is known as a *codon* and coding sequences in DNA are known as *exons* (for expressive sequences). Via messenger-RNA the exons specify how to build one of twenty amino acids. Only these twenty are found in all living organisms. Up to 600 amino acids are bolted together to form any one of the huge variety of molecules called proteins. The human genome contains information for around 100,000 proteins. A list of them would pretty much sum you up.

All organisms are little more than carefully arranged and maintained collections of proteins. Half the dry weight of a human being is protein. A T-bone steak is pretty much all protein. Whenever your body wants to do something it usually turns to a protein to do the job. They help build, maintain and repair the body and are especially important in muscle, cartilage and bones. Their influence is profound. All the structures in your body that are not made *of* protein were made *by* proteins.

Curiously though most of your DNA is not exons. The *introns* (for intervening sequences) in your cells don't seem to do anything useful. In fact in most higher organisms – the plants, animals, protozoa and fungi that have nuclei in their cells – most of this DNA does not code for proteins at all. It seems to be redundant: junk. Like the polystyrene peanuts used to protect fragile objects being posted, it just seems to be packing. Estimates vary but in the human genome some scientists think that only 3 per cent of the length of the DNA helix is actually used to code for proteins. Many think that 90 per cent of it is not doing anything immediately obvious.

Controversy has surrounded the role that this junk DNA plays. Some scientists have said that it really is nothing but junk and has no role to play. Others suspect that it plays a role, even if they don't know what it is. They liken it to a language we cannot read yet and are reluctant to say that it will never be understood. It might look like gibberish but with the right translator it might read like a Homeric epic.

To a non-English speaker, the sentence 'The wheelwright on my right is writing to say he thinks it his right to practise pagan rites. Alright?' would seem to be nothing more than the regular repetition of the same word. Yet to a native speaker it makes perfect, if somewhat contorted, sense. The same might be true of junk DNA: it might be written in the equivalent of a language we can't speak yet. As yet it is anyone's guess what information it holds and what it is trying to say.

Some parts of the sequence, called *telomeres*, count the number of times that some cells have divided. Every time the cell divides the telomeres shrink slightly. There is a school of thought that your cells can only divide a certain number of times before they die. At the time of writing this a very controversial theory. Other parts of the sequence play a different role. Some of them, called *centromeres*, are anchor points used when the cell is dividing. That still leaves 85 per cent or so with no discernible role. Some have likened this to having a 3000-page cookery book that has less than 100 recipes in the entire volume.[27]

At first glance the sequence of bases don't seem to make any sense. A few look like they were once coding regions but have become corrupted and have been abandoned like a broken bicycle. Often huge sections are simple repeats of two bases or stuttering sequences of nucleotides. But now some steps are being made towards understanding what is going on with introns. Curiosity on the part of Boston University physicist Gene Stanley has established that there is a pattern to the distribution of these introns. It is a pattern that is intriguing in its implications.

Stanley is one of the world experts on Universality. He wrote one of the first papers on the subject[28] before the field got its name and the ubiquity of the principle was recognized. He has spent a good proportion of his long professional career documenting the places that the principle of Universality applies. Even he is astonished at the sheer variety of places that it pops up. He has found it at work almost everywhere he has looked.

A Walk in the Helix

To reveal this pattern Stanley and his colleagues used a technique known as a random walk to plot the distribution of bases in the sequences between coding regions of DNA. A random walk is a simple but powerful way of revealing the history or spread of any series of events. In this case the random walk produces a graphical representation of the sequence of bases in a DNA string that is easier to understand and study than just a list of As, Cs, Gs and Ts.

As the name of this technique suggests it involves creating a metaphorical path or stairway for someone to walk along. Each step up and down along the path is determined by, in this instance, the bases that are encountered. Each base represents one step. The walk is intended to capture the fluctuating concentrations of pyrimidines and purines in the non-coding regions. It is a little like turning a flat map into a rugged three-dimensional model so that a blind person could get a feel for the landscape. Only in this case it is not a landscape that we are exploring but evolution. It is possible that the remnants of genes and other detritus in the non-coding sections of DNA are like a diary of life that records key events in that organism's developmental history – events that happened before an individual was born.

Stanley chose to have the walker step 'down' when it encountered a pyrimidine (cytosine or thymine) and 'up' if a purine (adenine or guanine) was in the way. A sequence of AAACG would involve three steps up, one down, and then another up. The walks produced by this technique are extremely jagged and have the look of self-similar structures. This look has been confirmed by analysis of the random walk. Stanley and his colleagues carried out random walk analyses on sequences of DNA thousands of base pairs long. They compared random walks from 33,301 coding and 29,453 non-coding sequences (each longer than 512 base pairs) from the 1996 GenBank database. GenBank is a repository of all publicly available genomic information. Other scientists have found correlations extending over the range of entire chromosomes.[29]

No correlations were found in the regions of the DNA that code for proteins. They seem to be working independently. However, long-range correlations were found in the non-coding regions. Sometimes these correlations stretch over vast sections of the DNA string. This means that the sequence of bases in one region has been influenced in some way or at some time by bases far from it. This has obvious affinities with systems in the critical regime. One of the signatures of this regime is that correlations can extend over enormous, effectively infinite, distances. Certainly at some point in the past, DNA was existing in a critical state. Certainly some coordination of effort has been going on but, as yet, no one is sure why and to what end.[30] It might be that this is evidence of life remembering what it is. The long-range correlations show that while everything else is changing something is preserved, some core information is being maintained. It might be that life is so precarious that we constantly have to tell ourselves who and what we are.

Further work by Stanley and his colleagues suggests what other ends this coordination might be working towards. Instead of tackling a huge section of DNA Stanley and his colleagues looked at just one well-known sequence and how it has changed across eight different species[31] right from yeast through worms and chickens to humans.

The sequence they picked is the one that makes the myosin heavy chain (MHC) protein. They picked it because it has been analysed across many species and this made the comparisons easier. The organisms they picked were a yeast called *Saccharomyces cerevisiae*, an amoeba called *Acanthamoeba castellanii*, the black-bellied honey-lover fruit fly or *Drosophila melanogaster*, four species of the worm *Caenorhabditis elegans*, a rat, a chicken and a human.

MHC is a protein used in skeletal muscles and it helps the fibres of this tissue slide over each other as an arm or leg is worked. The MHC molecule is huge. It is made up of around 2,000 or so amino acids. Most large proteins are made up of no more than a couple of hundred. As a result the sections of DNA involved in the making of this protein are correspondingly large. This makes them well suited to statistical analysis. They are also dotted with non-coding sequences of DNA. It is the distribution of bases in these that Stanley and his colleagues have studied.

A random walk through the MHC sequences for eight organisms found similar long-range correlations to those found by Voss, Munson and Stanley in their earlier work. Again, as the organism became more complex, the fractal complexity of the non-coding regions also increased with the evolution from yeast to humans. To gain an insight into what might be going on Stanley and his team produced a simple model that tried to capture what might have been happening over those aeons of evolution.

The model simulated the conversion of original coding regions into non-coding introns. Many scientists speculate that the non-coding regions are simply junkyards containing genetic machinery that has worn out and can no longer be used. Stanley set out to find out if this was true. The model assumes that RNA molecules were the precursors of all living things and that genetic information was originally encoded as a messenger-RNA (mRNA) molecule. In cells mRNA is the faithful hand-maiden of DNA. The mRNA reads the codons to find out which amino acid they have to make and then go off and make it. Stanley is assuming, and he is not alone in this, that DNA is just a storehouse of important information that cells use and mRNA reads to sustain themselves. The simulation tried to model the process by which the mRNA suggestions are turned into DNA sequences, which are then modified by mutation and insertion of non-coding material. Invading retroviruses, in effect diseases, can be responsible for inserting foreign sequences into DNA.

The simulation began with a sequence of DNA around 30,000 bases long that has a similar distribution of purines and pyrimidines to those in coding regions. At every step of the simulation a random point in the sequence is chosen and a length of DNA is cut from it. The length of this sequence is chosen according to another power law that is known to govern the size of loops generated by DNA. This section is then inserted at another random point in the sequence. There is also a 50:50 chance that some of the purines or pyrimidines in this subsequence will be swapped for their complements. There is an even smaller chance that the sequence being moved will be swapped for another random sequence that has the same percentage of purines/pyrimidines as the original long sequence.

When the simulation was run it produced distributions of purines/pyrimidines in non-coding regions that were almost identical to those found in simple living organisms like yeast. As Stanley and his colleagues ran more and more iterations, effectively mimicking the sweep of evolution, they saw the distribution of bases move in much the same way that they had seen when investigating how the MHC gene sequence has grown and changed in different organisms.

The Edge of Creation

What this implies is that, as Langton thinks, life is evolving at an edge. The processes of cutting and moving sequences of DNA tend to randomize the arrangement of bases in a strand, scattering the coding sequences and rendering them impotent. The reinsertion of sequences that possess the same distribution of the different bases helps the string combat this damage. It helps repair the structures and keep the coding regions functioning. The team ran the model without this repair mechanism and saw the DNA sequence gradually become randomized until there were no relations between any sequences.

When these competing forces – one trying to rip DNA apart and the other to restore it – are competing, the system exhibits power law correlations. The relationships between regions become stretched, chemical cycles that help maintain an organism stutter or slow down, sometimes forgotten but more often revived, as the DNA works to keep itself intact. It is reasonable to assume that if this antagonistic dynamic is at work in one sequence then it may be happening throughout all the non-coding regions of DNA. In these segments, just as in a real junkyard, you would see the shattered remnants of ageing coding sequences instead of recognizable fragments of machines such as wheels, pawls or gearboxes. On this scale life is very much a struggle.

Others suggest a different role for the non-coding sequences. Stuart Kauffman thinks that the self-similar patterns might be telling the cells what job they are doing. Kauffman believes that genes only know where they are because of the job they do. No genetic switches are flicked to tell a cell where or what it is. Instead, says Kauffman, like an apprentice in a factory cells learn their role as a body develops. They learn what it is to be a liver or spleen cell. Like a workaholic apprentice the cell defines itself by what it is employed to do.

Fractal patterns preserve order at all scales, even across time. Perhaps the memory effect of the distribution is a constant reminder to the cells of their place in the body and what they are supposed to be doing and which proteins they are supposed to be building.

Certainly it is known that DNA has not evolved randomly. Lawrence Hurst, an evolutionary biologist from the University of Bath, and Steven Freeland have shown that natural selection has been at work on DNA to make it among the best of mechanisms for preserving information.[32] The pair found that no more than one in a million randomly generated codes could do as good a job at preserving information as DNA. Further they found that single-letter changes to codons, which specify which amino acid to build, make cellular machinery build one with a chemically similar structure, thus limiting the damage. Perhaps the long-range correlations are the evidence for evolutionary forces that have shaped DNA into the copying machine that it has become.

The workings of Universality do not stop with RNA and DNA. It looks like Universality might also help structure the proteins specified in DNA and made by RNA. Proteins are found in a bewildering array of sizes, shapes and configurations. In the bodies of living organisms they are used to build structures, act as catalysts and transmit information. The hydrogen bond networks in water directly sculpt the form of a protein. The water-hating parts of the protein typically are on the inside of the folded structure with the water-loving elements bathing on the outside. Proteins are curious chemicals that seem to have the properties of both liquids and solids. The job a protein such as HMC does is determined by its shape. However, proteins are not rigid: Parts of them can undergo significant changes, while the other parts retain their structure and function. They assume their structure in a narrow temperature regime, above and below which the structure unfolds and loss of function follows the deliquescence of form. This might help explain why, temperature-wise, all life exists on a narrow boundary. If we get too hot or cold the protein machinery in our cells stops working.

Beyond self-sustaining chemical complexes, RNA, DNA, amino acids and proteins lies life itself. The prokaryotic bacteria that were masters of the planet for 2,000 millennia are little more than these five things put together in a flexible bag, and when you get enough of these bacteria together into colonies they tend to organize themselves into fractal patterns. The complexity of the pattern is determined by the distribution of food-stuffs in the medium in which the bugs find themselves. When nutrients are sparse the bacteria tend to be longer and the pattern more sparse, but when there is a glut the pattern is much denser and instead of being just a low-lying mat becomes a three-dimensional structure. Whatever the density of the patterns there is no doubt that they are self-similar. Some of the most ancient evidence for bacterial life are called *stromatolites*. These are ancient microbial mats and the nearest things to cities that bacteria create. And Universality was living along-side them.

Although the evidence is patchy in places and merely suggestive in others, it does look like the stirrings of life owe a debt to Universality. At almost every level, from the chemical through the proto-organic to DNA and beyond, the dynamics of this principle are at work. The origins of life might owe quite a debt to these forces. It certainly seems to have been contributing right from the start, and that was very much just the beginning of its influence.

Chapter Four

The Rhythm of Life

It may be that the laws of energy, and all the properties of matter, and all the chemistry of all the colloids are as powerless to explain the body as they are impotent to comprehend the soul. For my part, I think it is not so.

On Growth and Form, D'Arcy Thompson

It is beyond doubt that our body is a moulded river.

'Physical Observation No. 248', Novalis

The Heart of U

Your heart can sing many different songs. And when it changes its tune it is for reasons that your head will definitely understand and, perhaps, share. Most of these songs are familiar and you hear them every other moment of every day. Some are reserved for special occasions, and there are a few that only the dying will hear. These last rhythms make themselves felt when the heart is winding down, bringing a life to a close or racing pell-mell toward a cardiac arrest. Now, all these songs are gathering a larger audience, but not because heart disease is becoming more widespread. All the people listening are healthy and vigorous scientists with years of life left in them. They are collecting and studying these sounds because in the beating of a man's heart they have heard the pattern repeated time and time again in the larger world, and you have read about it almost as many times in the pages of this book. The song the heart sings is one of Universality.

The bumping beat of our heart is the first and last noise we will hear, the personal soundtrack that we carry everywhere with us. Usually it is a reassuring presence, a polite and permanent guest that is reluctant to intrude. Only in the throes of passion, exercise or fear does it threaten to overwhelm us. The rest of the time we are happy to ignore its insistent tempo but if we listened to it we would find out how subtle, complex and compelling its beat can be.

If you take your pulse now while you are resting and reading you should find that your heart beats between 70 and 80 times a minute. The average is around 72. The resting rate of a mammal's heartbeat is inversely related to its body weight. So, the larger the animal, the slower its heart beats. An elephant heart beats about 30 times per minute and that of a shrew hammers along at 800 beats per minute. If you live to a ripe old age your heart will beat up to 3 billion times in your life. A foetus's heart starts beating around two months after conception. Every minute the heart pumps your entire blood supply around your body. There is nothing so loyal and constant to you as your heart.

If you were very careful and listened very closely to your heart you might notice that the pauses between beats do not fall regularly. Even when you are not doing any exercise the pause between beats varies. Sometimes the heart pauses momentarily before beginning a new thump, before hurrying to start the next pulse. All this is perfectly healthy.

The heart is really just a big muscle and would tire quickly if it performed exactly the same movement every time it pumped blood. To you it might sound like it is beating out a regular rhythm but it is not a metronome. Anyone who takes regular exercise will know how quickly muscles can tire if forced to repeat one movement time and time again.

The heart avoids fatigue by varying the speed and rate at which it clenches. The changes in rhythm are too subtle for human hearing to pick up but they make all the difference to the cardiac muscles. The coordinating centre of the heart is known as the sinoatrial node: the pacemaker. Here begin the pulses of electrical activity that make the heart muscle cells contract as one; but the call to clench, in the form of a wave of electrical activity, does not always come at the same interval. The signal varies in strength and urgency so that almost every command is different. This means the heart never makes the same movement from moment to moment. Movements are repeated but on such long time scales that a healthy heart never tires.

Healthy competition between the sympathetic and parasympathetic nervous systems creates this erratic rhythm. These two control systems are the cheerleaders of the body and try to get all the organs to respond to their calls. The parasympathetic system is constantly trying to relax the body and slow down the firing rate of that coordinating node. By contrast the sympathetic nervous system is trying to goad the body and get it ready for action. The outcome of this push-me-pull-you competition is the complex and unpredictable interbeat rhythm.

The erratic beat of the heart also helps it adapt to the different loads placed upon it by the body carrying it around. If it simply assumed a single regular rhythm it would find it difficult to change its tempo quickly, but the erratic rhythm means that it is effectively beating at all speeds all the time. It has no expectations of what is normal and so is happy to do whatever the body demands and speed up or slow down as necessary. If, as looks likely, the heartbeat exists at a critical point then like all such systems it will be very sensitive to small changes. As a result it will be able to react very quickly to the demands that the body is placing on it.

Only now are we beginning to appreciate what a well-designed pump it is. The complexity of the interbeat intervals has only come to light in the last ten to fifteen years. Prior conceptions of a regular sinus rhythm have had to be abandoned, but then the history of the heart is riddled with misconceptions. Aristotle thought that it was the seat of the intellect because a person's voice issued from their chest. Ancient Egyptians saw it as the repository of the soul and truth in a man.

Even William Harvey's interest in the heart stemmed from a muddle-headed opinion about the body. Like many men of his day Harvey assumed that the body was a reflection of the universe, only much smaller – a microcosm of the macrocosm. He thought that because there was a single ruler to the universe there must be the same at work in the body. Remembering the work of Copernicus, who proved the planets revolved around the sun and not vice versa, Harvey decided to investigate the organs at the centre of the body. He got to the right organ but for entirely the wrong reasons.

Whatever motivated Harvey his contribution to science is sure. In 1628 he produced *An Anatomical Treatise on the Motion of the Heart and Blood in Animals*, which marked the beginnings of physiology proper and must rank as one of the most influential scientific works ever written. Nevertheless, it wasn't an easy treatise to write and Harvey had to exploit his connections as physician to King Charles I to get enough samples to anatomize and perform his physiological experiments.

Now it turns out that the heart is much more complex than Harvey ever could have imagined. It shares many of the properties of systems that are far from equilibrium: systems that are near a critical point.

Life Rules

Electrocardiograph analyses of the between-beat intervals over long periods of time, days in some cases, has shown that these delays possess a fractal form. They are remarkably similar to the flicker noise that is characteristic of many systems displaying Universality. If you plot a graph of the interbeat intervals and look at it using different time scales it looks irregular and wrinkly. In fact if you aren't given a scale you wouldn't know whether you were looking at a record covering seconds, minutes or hours. The distribution is self-similar. This is true at least of healthy hearts, even those of cosmonauts who are in orbit. Research has shown that the interbeat intervals of old people or those suffering from congestive heart failure have a different non-fractal pattern.

The fractal pattern is characteristic of a healthy heart because it shows that the organ is avoiding becoming tired by irregularly varying its pulsing pattern. Fractals are evidence of long-term correlations at work and help the heart remember what it did tens or hundreds of beats ago to ensure it does not fall into a repetitive pattern and get tired.

By contrast the hearts of patients suffering heart disease do not possess these fractal patterns. This suggests that their hearts are not as responsive or flexible as those of healthy individuals. People with sleep apnoea show similarly inflexible interbeat patterns. In this condition the tongue of a sufferer blocks the flow of air to the lungs hundreds of times per night. The lack of oxygen wakes the sufferer with a snort and is known to cause chronic fatigue; it is also linked to memory loss, high blood pressure, stroke and heart attacks. It is perhaps not surprising that sleep apnoea leads to all these health problems because it is thought that problems with the autonomic nervous system that controls heart rate and breathing are the root cause of the condition.

Work is now going on to see if this difference

between the hearts of healthy and poorly people can be used to diagnose cardiac problems before they make themselves known in more dramatic fashion by bringing on a heart attack. The problem is that at the moment records of interbeat intervals have to be collected over a period of hours before the fractal pattern can be determined. However, the technique could be useful in monitoring patients who are under observation and confined to a hospital bed or those who can walk and don't mind taking their heart monitor with them. However, scientists such as Ary Goldberger who are researching this subject are refining their techniques to make faster diagnoses of healthy and poorly patients. In industrialized nations, where heart attacks kill one in four people, a fast method of diagnosis could save a lot of lives.

This research might also produce gentler artificial pacemakers for unhealthy hearts. Often people who have suffered some form of heart beat abnormality, called fibrillation, have a device implanted in their chest that can shock the heart back to normal when it goes awry again. In fibrillation the heart goes into a frenzy of beating and cannot regain its irregular healthy rhythm. Existing defibrillators are relatively crude devices. They only take action when the heart is actually starting to spasm. They need to deliver a big shock to get the heart back to its normal rhythm. This big punch is very painful and it can take several hours for people to fully recover from it.

Now Bill Ditto from Georgia Tech in Atlanta is working on a defibrillator that gets to know the fractal beat of your heart and delivers tiny shocks to steer it away from fibrillation. The smart device can spot the danger signs when the heart is straying from its irregular rhythm. With a few small well-placed shocks Ditto believes it will become possible to prevent the heart even entering the runaway fibrillating state. Such devices would use far less energy and be much less painful than existing pacemakers.

Ditto's work on defibrillators might also help other very sick patients because the heart is not the only organ that moves to a fractal rhythm. The pattern of our breathing follows a self-similar pattern. However, when patients who cannot breathe for themselves are put on life-support machines they effectively take the same breath every time they breathe in and out. Artificial respirators are usually set at one breathing rate and push the same volume of air into a patient at every inspiration. This can cause the tiny alveoli sacs deep in the lungs to collapse, making it harder for an already very poorly patient to breathe and recover. Work is being done on a respirator that varies the amount of air it pushes into a patient's lungs using a fractal distribution similar to that seen in healthy working lungs.[1]

The implications of this work are far reaching and only just beginning to be appreciated. Much of modern medicine assumes that a healthy body is an ordered body: hence the description of an ailing patient as someone suffering a 'disorder'. It teaches that homeostatic mechanisms in the body work to bring it back to equilibrium if it is pushed off balance by illness or injury. As a result diagnosis is a search for disorder and treatment all about restoring balance.

This work also gives new insights into what counts as ageing. Older bodies tend to be less adaptable than younger ones and have a more limited fractal repertoire. As a result the responses of their bodies to illness or trauma are not as creative. Many medical researchers are now investigating this loss of flexibility and looking into ways of restoring it or delaying its disappearance.

As finer mechanisms for measuring heartbeats are showing, some organs exist in a dynamic state, far from equilibrium. Moreover, it is not just the heart and lungs that show the characteristically irregular, but not random, pattern. The population of white blood cells in the body goes through similar fluctuations. Even brainwaves show this erratic pattern. In all these situations the body is maintaining itself at a critical point, poised and able to react swiftly to any change. This is much healthier than a body that produces the same regular output and reacts linearly to increased stress. It seems to be the case that systems disordered in the right way are highly adaptable and probably far healthier than those repeating the same motion over and over again.

An organism that is too rigid in its responses will quickly get caught and wiped out. One that overreacts will spend its time coping with a succession of crises rather than getting on with living. In between these extremes is the critical region that is a good trade-off between adaptability and rigidity. The tension between these two forces keeps us alive, hale and hearty.

Cardiac Classics

All this talk of hearts singing songs is not just idle whimsy or literary fancy. There is no doubt that, with a little help, the heart can play tunes. In 1998 composer Zach Davids and his father, cardiologist Ary Goldberger, released a CD of 20 tracks called, perhaps inevitably, *Heartsongs*. Goldberger works at the Beth Israel Deaconess Medical Center in Boston and is the man who first found fractals in the heartbeat. Each song was based on the data gathered from different people's heartbeat patterns. Around 10,000 heartbeats were used for each song.

The precise interval between beats was measured for each heart, producing a graph of the moment to moment changes in heart rate against time. The interbeat delays were then converted into whole numbers and used to regulate the melody in each piece. The intervals determined the frequency and duration of each note. Composer Davids then added chords and rhythm over the top of the melody to make the whole thing more musical and pleasing to hear.[2]

Music played to the tempo of the heart has a subtly pleasing rhythm. It speeds up and slows down, meandering unpredictably, and never really settling down. If you listen to the music you find that it does have a curious kind of resonance. It reminds you of something but you're not sure what. My reaction to the music is not perhaps as pure as it might be. I know what went into the creation of these tunes and maybe this unconsciously determined my critical reaction.

You can hear raw heart music if you want but it does mean taking a trip to the Boston Museum of Science. An interactive section of the Dance of Chance exhibition allows visitors to record their own heart data and have it instantly converted into music. Many people find that this heart music is faintly familiar and not just because their own heart is pumping out a similar rhythm. There is some evidence that we enjoy some music only because it is an echo of the rhythm thumping at our core.

In 1941 Russian-American music theorist and obsessive Joseph Schillinger published a twelve-volume work that attempted to codify written music and its composition according to his own ideas. Although the work contains some decidedly wacky ideas it did anticipate a lot of work that has now become part of complexity theory. Years before Mandelbrot coined the term 'fractal' and was working out how to deal with intermediate dimensions, Schillinger was writing about the different results you would get if you measured the North American coastline on astronomical, topographical or microscopic scales. Much of the rest of Schillinger's *magnum opus* was concerned with the rhythm of written music and the patterns that composers intuitively choose. Although the terminology was not yet established Schillinger found that the rhythms were fractal.

In 1978 Richard Voss from the IBM Watson Research Center and John Clarke from the University of California carried out a slightly different analysis of music.[3] Instead of looking at the structure of the music as it was written the pair looked at how it sounded.

The five scientists who have done most to establish
Universality and show how it relates to the real world.
Leo Kadanoff *(top left)* was the first to realise how
complex were the dynamics of systems at critical
points. Kenneth G. Wilson *(top centre)* was awarded
a Nobel prize for working out the formidable
mathematics of this closed world. Benoit Mandelbrot
(top right) has spent his life chasing fractals - the
signature of the critical region. Gene Stanley *(bottom
left)* has spent much of his career spotting other
systems exhibiting the hallmarks of Universality.
Per Bak *(bottom right)* reawakened interest in the
field and has helped define its limits.

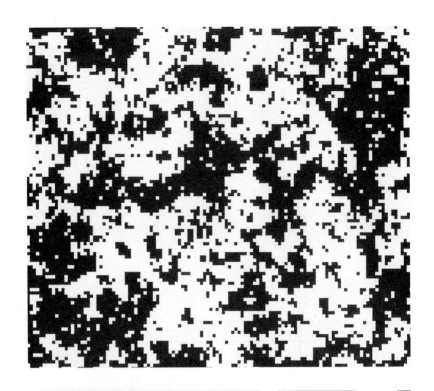

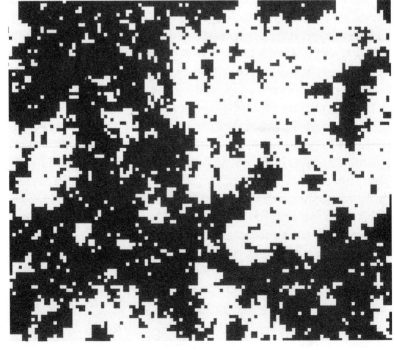

Simulations of a section of an iron bar magnet below, at and above the critical temperature. Below the critical temperature *(opposite page, top)* of this system (770°C) the bar is magnetic because the tiny magnetic grains it is made of share the same orientation. As it is heated thermal energy causes some of them to flip and the strength of its magnetic field wanes. At the critical point *(opposite page, bottom)*, clouds of magnetism ripple across the bar as it is kept at the same temperature. Just enough energy is being added to sustain it in this indecisive state. The properties of the system have ceased to matter and its organisation has become key. If successively smaller sections were magnified they would show the same fractal distribution. The dynamics of one region are characteristic of the whole. Above the critical temperature *(below)* the tiny grains are constantly kicked around by the energy being added and the system descends into random, white noise.

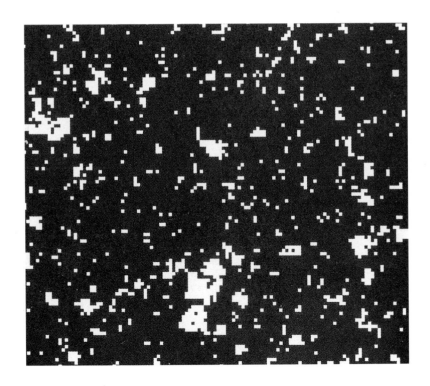

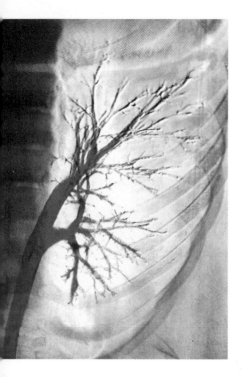

Left and below: The fractal filigree found in lungs and leaves is a highly efficient distribution system. They have evolved for the same reason and do the same job.

Right: Fractals are ubiquitous throughout the natural world. Here an algal bloom forms a fractal pattern as it grows in the Baltic.

Below right: When wildebeest migrate they do so in herds displaying all the signs of Universality. The complex structure of the herd helps to co-ordinate its direction and to spread information about predators quickly.

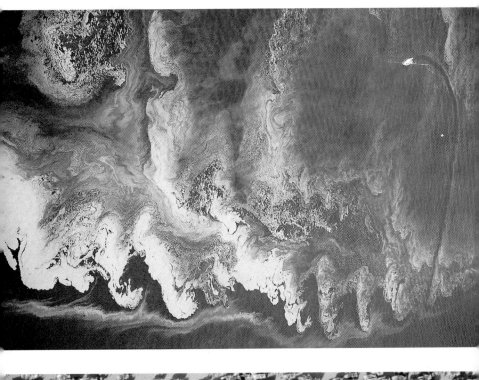

Over aeons landscapes have been sculpted by the forces that the theory of Universality lays bare. Here snow reveals the structure of mountains in Utah.

Some fractal structures are only short-lived. This image of Pacific waves shows how wind, water and tides can combine to produce a transient pattern that has all the hallmarks of Universality.

Frost on a window pane has an intricate structure
reminiscent of computer generated fractal patterns.

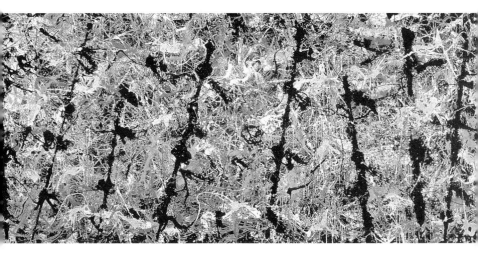

Unknowingly artist Jackson Pollock invested many
of his most stirring works with a fractal pattern.

An image of a face generated using fractal algorithms. Some researchers are asking whether what we consider to be beautiful is influenced by the ease with which we can perceive it. Smooth, fractal faces are easier to perceive than craggy, wrinkled visages.

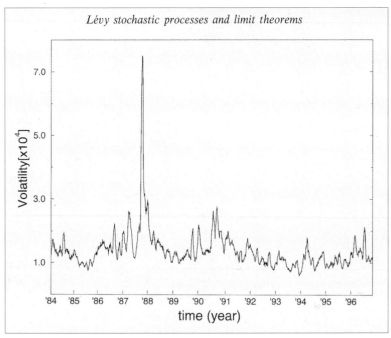

Lévy stochastic processes and limit theorems

A graph showing the range of fluctuations in the Standard and Poor 500 Index over 13 years. Close analysis of the changes has revealed a fractal structure driven by the buying and selling decisions of millions of traders.

They analysed the audio signal piped to the speakers as the music was played on a stereo or radio. To get an idea of the audio power of the music they measured how much electricity was being used to drive the speakers as the music was being played. The audio power of a piece of music is proportional to the power delivered to the speakers to reproduce its constituent sounds. In the music they found $1/f$ noise – just as has been seen in so many systems displaying the characteristics of Universality.

Almost all music seems to be played in the key of $1/f$. Voss and Clarke analysed Bach's First Brandenburg Concerto and a recording of Scott Joplin piano rags. Both had a $1/f$ signature. The pair also looked at music from rock, classical and news talk radio stations. In all three $1/f$ noise was present. Voss has tried composing music using $1/f$ noise from the natural world. He even composed one piece derived from the flood levels of the River Nile. In April 1978 the *Scientific American* columnist Martin Gardner proposed a way of generating $1/f$ music for yourself using nothing more than three six-sided dice of different colours, a piano, some coloured pens and a chart.[4]

In fact there is a huge amount of classical music that is pitted with fractals. Many classical pieces are about variations on a theme in which a particular musical phrase is repeated with slight changes. In geometry it is possible to produce a fractal pattern by repeating the same step over and over again. The fractal shape known as a Sierpiński carpet is made in this way. The shape is named after the prolific Polish mathematician Wacław

Sierpiński. Imagine a largish square, say one big enough to fit on a piece of A4 paper. Now divide it into nine equal but smaller squares and remove the middle one. Do the same for the remaining eight squares and keep going. The simple rule eventually produces a complex shape that hovers between one and two dimensions. No matter where you look in the carpet the original pattern is produced on any and all scales. There are many pieces of classical music which work the same way. They take a starting phrase and work on it endlessly and delight-fully.[5] The same distribution of notes is repeated on different scales and times. If you analyse the scores of these pieces, you will see the same pattern repeated just like in the Sierpiński carpet, albeit in a myriad of forms.

Johann Sebastian Bach in his Toccata and Fugue in E Minor plays around with a trio of notes in a strikingly self-similar way. Bach's monumental Art of Fugue (*Die Kunst der Fugue*) starts with four fugues. Two form the main theme of the piece and the other two are the same theme but back to front. Then there are counterfugues that turn the original upside down and are then mixed with the original. It also contains double fugues, triple fugues and even mirror fugues. Part of one composition in this work becomes an element of another and each is endlessly reflected. This Art of Fugue was Bach's last major composition but he never finished it. The last piece in the composition is a triple fugue that ends just as it starts to musically explore the letters in his name. The 'B' is a B-flat and 'H' is a B-natural in German musical notation.

The same kind of pattern exploration, but not with

the same colossal range, can be found in Bach's First Brandenburg Concerto or Beethoven's Fifth Symphony, which is built around only four notes. Perhaps this music is so stirring because it chimes with the subliminal rhythm that echoes through our body every day. This might also help explain the change in Beethoven's work brought about by his deafness. Typically the life and work of this great composer is divided into three parts. An early imitative phase, then a period characterized by expansive heroic works and a final concentrated late style. Many of Beethoven's biographers believe that this latter change was driven by his increasing deafness and the despair he experienced as it gradually cut him off from hearing music just as he was enjoying great success.

This despair also drove a period of immense creativity. Part of this was due to the fact that being deaf made it impossible for Beethoven to perform in public. As a result he had more time to compose. It might be the case that Beethoven produced his best work at the end of his life when his increasing deafness cut him off from the world and left him free to listen to himself.

He could do this because as with all composers music coursed through him at all times. He was so steeped in it that novel workings and compositions bubbled up through him constantly. In a letter to a friend he wrote: 'I live entirely in my music, and hardly have I completed one composition when I have already begun another.' Towards the end of his life he filled hundreds of musical sketchbooks with half-worked-out pieces and variations on themes that had occurred to him. He wrote to one of his pupils: 'What is in my heart must come out and so I write it down.'

It was during this time that he produced some of his most dense and challenging work. This music is so difficult to play perhaps because Beethoven couldn't listen to it being played. Instead he worked with what he heard inside him unfettered by the limited dexterity of his hands on the piano or the ability of an orchestra to perform the composition. This late period saw the composition of such works as the Eroica, the Hammerklavier sonata, the Diabelli variations and the Ninth Symphony.

In a very real sense the music is in tune with us and our surroundings. When we hear it we are recognizing ourselves and the ringing echo of Universality sounding through the world.

Fractal Faces

Bach, Beethoven and most other classical composers do not set out to create music with a fractal structure. It just happened that way. Theme and variation was the fashion in music for a long time and even those who set new standards, such as Bach and Beethoven, and moved music on, did so by bending those rules almost out of shape. Despite this the origins of their music are still evident. They produced this music for reasons of musical fashion and personality, but largely because to them it just sounded right.

Now many composers explicitly use fractals to generate music. London-based computer programmer and amateur musician Phil Thompson released an album of fractal music called *Organised Chaos* in October 1998.[6] One of the tracks on the album was based on the Mandelbrot set – a fractal pattern named after the man who started it all off. The Mandelbrot set is a fearsomely complicated pattern. When drawn by a sufficiently powerful computer it is possible to drill down through its levels and encounter seemingly endless vistas of complexity. It is a little like standing on the edge of the Milky Way, and peering down towards our Solar System, towards the earth, looking through the clouds and being able to read over my shoulder as I type. Generating this complexity, however, only takes a few lines of computer code. By repeating the same set of instructions an awesomely deep pattern emerges. To reflect the intricacy of this pattern and its infinite levels Thompson called the piece 'A Season in Hell'. The album had its debut on BBC Radio 4 and generated a huge amount of interest. When it was first aired Ian Stewart, professor of mathematics at the University of Warwick and no mean fractal expert, said: 'What amazes me with this kind of music is that it sounds much better than you would expect.'[7]

Then there is David Clark Little, a composer and concert harpsichordist based in Amsterdam, who has used fractal programs to compose music for the last 10 years. California-based Forrest Fang has even composed an opera some of the songs in which were generated using fractals. The opera was performed for the first time in late 1999. Fang was something of a late convert to fractals and before he discovered them made his living as a session musician playing ethnic string and percussion instruments such as Turkish lutes and Filipino gongs. Now he is a full-time fractal performer.

The implications of this fractal form go far beyond music. In fact fractals might be the key to our entire aesthetic sensibilities. At the turn of the century American mathematician George David Birkhoff put forward a theory about what makes a work of art, in any form, pleasing and interesting. He said that it should be neither too regular and predictable nor pack in too many surprises. Apparently we like to be stimulated but not overly so.[8] This theory was lent some weight by the work of Voss and Clarke. The pair tested music generated with different power spectra. They created music from the frequency dependencies found in white noise and in Brownian motion. People preferred the noise derived from the $1/f$ noise. The listeners found the white noise too random and the Brownian motion music too regular.[9]

Fractals also seem to play a part in what we consider beautiful. Although culture obviously plays a role in determining what we find attractive it does not account for everything. David Perrett, a neuropsychologist from the University of St Andrews in Fife, has tested cross-cultural ideas of attractiveness using composite faces. These are pictures built up by overlaying two or more other facial images that people have said they find attractive. He found that all of us find the same things attractive.[10]

Women like features that are indicative of sexual maturity, such as a strong jaw line and prominent chin. Also preferred are small noses and large eyes. Men have a much longer list of attractive features. Males like large pupils and widely spaced eyes, high cheekbones, a small chin and upper lip, and a generous mouth. Younger women tend to have these features, so it is speculated that men are recognizing the signs of higher fertility. Men also like glossy hair and smooth skin, both of which are the result of high oestrogen levels and again are traits associated with the young and highly fertile. The composite images of what people find attractive have been tested on people from many different cultures and racial backgrounds and there is general agreement about what is attractive and beautiful: mature men and young-looking women.

However, there may be another force at work too. Jürgen Schmidhuber of the IDSIA Institute in Switzerland thinks that beauty has as much to do with how easy we find it to mentally process a face as it does with the traits we look for in a mate. Schmidhuber has carried out experiments with the faces used in previous experiments to study cross-cultural ideas of attractiveness. To the possible choices he has added some that have been subtly altered. These other faces were generated using fractal transformations. Schmidhuber claims that among his subjects there was a marked preference for the fractal faces. He puts this down to the face that fractals make it easier for the sensory systems in our brain to encode the information they are seeing. The self-similarity makes it a cinch to see. The result is that we find faces that are easier to process more beautiful than those that take more mental horsepower.

Leonardo da Vinci was known to use mathematical

rules to generate faces in some of his work. Even before that he speculated on the internal workings of a tree and how it must be shaped in order to distribute sap around all the branches. It is an observation that has stood the test of time. Perhaps the *Mona Lisa* is so beguiling because it is so easy to turn her face into mental impulses and that we see a reflection of the wider world in her enigmatic smile.

It is an intriguing idea but, like much of the work being done in this field, it is at a very early stage. The preference for the fractal faces could have other explanations. The researchers who did the work on cross-cultural comparisons of attractiveness found that people preferred the composite images of people to the real images. This was because the blending of images removed freckles and other blemishes from the faces. Schmidhuber's images are blemish-free and might be thought more beautiful simply because they are flawless.[11]

Pipes and Plumbing

Fractals can be used to generate more than pretty faces. Large organisms are riddled with them. Many of the organs in our bodies and the structures connecting up all the bits have a fractal component. There are few structures in the heart that are not fractal in one way or another.

The arterial and venous trees that take blood to and from the heart are fractal, as is the branching of the bundles of cardiac muscle. The network of nerves that distributes electrical impulses to the ventricles of the heart, called the His–Purkinje network, is a fractal tree too. It reaches all parts so that they clench in unison at the right moment.

A self-similar geometry also underlies some of the mechanics of the way that the heart pumps blood. The mitrial and tricuspid valves are held in place by a fractal filigree of connective tissues called the *chordae tendineae*. The mitrial valve connects to the aorta and the tricuspid connects the chambers of the heart. Without such strong anchoring points the pumping heart would rip itself apart. The bundles of muscle within the heart also display a fractal geometry. The bunches of cardiac cells, called myocytes, are made up of bundles within bundles within bundles. The largest units are the bundles which themselves are made up of muscle fibres; these split down into smaller fibrils that are made up of the muscle proteins actin and myosin. And so it goes on.

While it has long been suspected that blood vessels and the bronchial branches were fractal it has only recently been established that this is the case. In early April 2000 Yonathan Shapir and Jacob Jorné of the University of Rochester showed that when nature is building such things it can form fractal structures.[12] Shapir and Jorné established that cyclical phenomena, encompassing such things as the growth of tissues as well as lawns and charge depletion in chemical batteries, follow the principle of Universality. It might be the case that our bodies are frozen fractal structures covered in a sheath that was itself laid down fractally. Shapir and

Jorné have established that tumours grow in a fractal pattern and it is likely that many of the organs in the body formed in a similar manner.

The success of this approach is evident because now diagnostic tools are being developed that can distinguish between healthy and abnormal tissue. The distinguishing factor is the fractal dimension of the tissue. We are riddled with these things.

Certainly many biologists are keen to redraw the evolutionary tree, which before now has been organized by similarities between skeletal and body shapes. The advance of molecular genetics has shown the real relationships between species, with cows closer to dolphins than horses and elephant shrews cousins of real elephants rather than other insectivores such as moles and hedgehogs.[13] Genes, not morphology, are the real basis of similarity. Universality may have helped direct the forms we take.

Moreover, this direction is not confined to us humans. Organs in other organisms are structured in the same way. The network of spiracles many insects have in place of lungs branch like fractals. It has even been claimed that many animal markings are the result of a self-repeating process. Certainly there is a resemblance between many of the patterns produced by cellular automata and those found on some cone shells. The patterns of the shells of a snail known as a tent olive (*olivia porphyria*) are easily reproduced with 2-D cellular automata. Snail shells are formed like fingernails and perhaps the pattern on each new layer forms in response to the design on the one that precedes it. There is no doubt that the shells of some creatures such as the nautilus do grow in a self-similar pattern. Universality seems to be the foreman directing the development of entire organisms.

The reason the fractals feature so strongly is because they are very efficient ways of distributing anything, be it oxygen or blood, over large spatially distributed systems such as the innards of you and me. We seem to be made of interwoven frozen fractals. In fact research by ecologists James Brown and Brian Enquist of the University of New Mexico in Albuquerque and physicist Geoffrey West of the Los Alamos National Laboratory has shown that a fractal geometry is the most efficient way of distributing what a body needs around that organism.[14] Certainly the fractal arrangement of blood vessels in your body manages to cram a lot into a small space. The work on Sierpiński carpets and gaskets revealed that. It is estimated that you have around 60,000 miles of pipes crammed into your body: that's more than twice around the world. If that is hard to grasp then think of your lungs which manage to cram a surface area as big as a tennis court into a space little bigger than the inside of a carrier bag.

Brown, Enquist and West found a relationship between the size of an animal and the complexity of the branches in lungs and blood vessels that holds true over the twenty-one orders of magnitude that organisms span – from the mouse to the whale. The fractals are so efficient a distribution system that they allow large organisms to need a far less complex network of veins and arteries than you would expect. A whale is 10 million times heavier than a mouse yet needs only 70 per cent more branches in its circulatory system to supply its body with oxygen and other nutrients.

A fractal form maximizes surfaces used to exchange such things as oxygen and it also guarantees relatively short journey times to the farthest reaches of a structure. Nothing else works as well. If we had to rely on oxygen just diffusing out of our lungs around our body it would take around three years for it just to reach our hands.

Nature has a habit of sticking with what works well, which probably explains the ubiquity of fractal branching mechanisms throughout the natural world. All the way from the veins in the back of your hand to the branches on a tree, they are the same tool for the same job.

Distance Doubts

Nevertheless, the existence of these fractals has been disputed. By their nature geometrical fractals extend over infinite orders of magnitude. With a decent computer it is possible to take a dizzying journey into some complex fractals and cross hundreds of orders of magnitude. There is no doubt that these patterns are fractal because over many different scales the same features are seen. Self-similarity is after all the signature of this phenomenon. The difference between 1 and 10 is one order of magnitude; the jump from 1 to 10 to 100 is two.

Yet in organisms self-similar structures have definite beginning and end points – they are by no means infinite. In the heart, for example, the fractal structures start with the myosin and actin protein chains in the muscle cells and end with the whole heart. At most this covers a few orders of magnitude. If we drew a Sierpiński carpet with only two iterations (equivalent to a couple of orders of magnitude) there would be nothing remarkable about it. The central square would be gone as would the middle square in all of the remaining eight squares. And that's it. It would look more like a design for a potato masher than it would suggest universal organizing principles.

To make matters worse the features that appear on different scales are not as well defined as those drawn by a computer working to a particular equation. In mathematically derived fractals exactly the same features are seen at all scales. In the real world shapes and structures do not possess the precision of geometrical figures. They are less precisely drawn, fuzzier, but no less distinct and their self-similarity is evident.

Some scientists have questioned whether the name 'fractal' should be given to these structures at all because often there is uncertainty about their reach. While a case can be made for some fractals in organisms and nature many are inferred from shaky data, sometimes less than one order of magnitude.

A row over exactly this point erupted in the journal *Science* in early January 1998. The argument was kicked off by a feature written by Israeli scientists David Avnir, Ofer Biham, Daniel Lidar and Ofer Malcai which presented the results of an informal survey of papers claiming evidence of fractal phenomena.[15] They looked at all the papers over a seven-year period submitted to

Physical Review Letters and all its sister journals. In all they looked at ninety-six papers that dealt with fractal phenomena in nature. The survey found that in most of the papers the fractals studied extended over less than two orders of magnitude, or decades. The average was 1.3 orders of magnitude. The researchers wrote: 'This limited range stands in stark contradiction to the public image of the status of experimental fractals.'[16] They asked if nature was really fractal after all.

Unsurprisingly the article drew a retort from Benoit Mandelbrot in the form of a letter to *Science* published later the same month. Mandelbrot partially agreed with Avnir and his co-workers that fractals are not everywhere. He says as much in the opening pages of *The Fractal Geometry of Nature.* He also admitted that many of the reports on the existence of fractals are the result of enthusiasm rather than hard evidence. However, he also pointed out that there are many, many situations in which the existence of fractals over many decades are not disputed. He mentioned finance, metal fracture data and the fractal bursts of transmission errors as particularly good examples. He pointed out that many of the situations in which short-range fractals were first found have, on subsequent research, turned out to extend over many orders of magnitude.[17] In the lungs, for example, there are over twenty different branches between the large bronchus at the top of the lungs and the tiny alveoli that it terminates in. In his typically mischievous style Mandelbrot also pointed out that at least one of the scientists carrying out the survey, David Avnir, had published papers revealing fractals that extended over less than one order of magnitude. Mandelbrot wondered if now the claims of those papers would be withdrawn.

The episode was a short-lived public spat and it would be wrong to blow it up as a rift among complexity theorists. However, the verbal wrangling reveals a fundamental problem with fractals in the real world and therefore for Universality. Just how ubiquitous is it?

Certainly a lot of bold claims have been made for Universality. Not least the name it was given back in the 1970s which, like chaos theory, is not the most helpful description that could have been bestowed upon it. It is obvious that Universal here does not mean all-encompassing. An awful lot of the events in the wider world can be described and explained with very mundane and well-established theories. Universality should not be taken as trying to usurp all other explanations. However, there are an equally large number of situations that Universality is providing insights into. Often many of these events were impossible to study and model before the methods of tackling the dynamics of critical points were established.

Gene Stanley, who has done more than most to further the study of Universality, says the choice finally comes down to the scientists doing the research. For himself he is not satisfied to label something a fractal unless he has at least two decades worth of data to dissect. Some scientists may be more demanding; others less so. Again, it has to be said, the study of Universality is in its infancy. There will be mistakes made and probably some bogus examples will turn up. As the data is weighed and the research is done it should become obvious where the legitimate claims are being made.

One, Two, Many

There are no lack of places for people to look for evidence of fractals and Universality. It appears that the dynamics of the critical point are at work within organisms on many different levels. On the smallest scale they help keep cell walls intact but also give them flexibility. As a body grows they help to determine structure and form. The fractal trees in the bodies of many organisms ensure that oxygen and other nutrients are passed around efficiently. Some of the organs linked by these networks are themselves constructed along fractal lines and some, such as the heart, lungs and liver, mark out a self-similar rhythm as they help keep someone or something alive.

It is becoming apparent that living organisms, as well as much of the inanimate world, are not represented in their infinite variety. Universality implies that there are very real constraints on the forms that organisms take. Straightforward physics, which takes a lot of theory to explain, provides those restraints. The behaviour of many organisms does not escape these restrictions either. The pull of Universality does not end where the skin stops. There is ample evidence that the actions of many organisms traces out a self-similar pattern.

Studies of the way people walk has revealed that we employ only a few of the possible number of gaits available to us to get around. Similar studies of animal stride patterns have revealed the same tendency to use fewer ways of walking than there are available. An animal with four legs has a lot of choices about the order in which its legs are moved. Yet only a subset of all these possibilities is used.

Walking is coordinated by lots of different systems in the body. Input comes from the eyes, the nervous system, the inner ear, the leg muscles themselves and the joints from the hips on down. It's a surprise that we ever get anywhere. The output of this interaction is walking and the pattern of strides it produces appears to follow a power law. Analysis of the differences in stride lengths of healthy walkers has revealed a self-similar pattern. Like all such systems this is a very adaptable arrangement and allows us to cope with different terrains and change speed very easily. Curiously the gait patterns of older people and those suffering motor-neurone conditions such as Huntington's disease did not show these long-range correlations.[18]

The movements of other organisms show similar patterns. The wandering albatross (*Diomedea exulans*) is the epitome of the solitary creature and it spends most of its fifty- to sixty-year life span on the wing. They are one of the few birds that die of old age. Young wandering albatrosses spend the first ten years of their life at sea, never coming to shore. While at sea they drink salt water and feed off squid, cuttlefish and the rubbish thrown from ships. They even sleep bobbing up and down on the waves. The birds only really come ashore to breed, and when they do they form huge colonies on some of the most remote islands in the world, places such as Tristan da Cunha. Their tameness on land led to them being called gooney birds or mollymawks (from the Dutch for 'foolish gull').[19] The wandering albatross is very well adapted to its peripatetic lifestyle and is built for staying aloft with the minimum of effort. The span of its wings from tip to tip can measure over 10 feet.

A study of a tagged wandering albatross called Mrs Gibson in 1996 showed that they can fly an average of 300 kilometres per day. Occasionally the birds have been known to cover distances of 1,000 kilometres in a twenty-four-hour period. Mrs Gibson plied the winds to fly over the sea off the Eastern coast of Australia in a series of arcs and covered distances of 30,000 kilometres or more over a period of nine weeks. The wandering albatross is such a good glider that Joachim Huyssen from the University of Pretoria in South Africa is using D. exulans as the inspiration for a gliding aircraft.

Albatrosses do not just follow the wind, they use it to get where they want to go. Gandimohan Viswanathan from the University of Boston has studied the foraging habits of albatrosses and found that there is a self-similar pattern to their wanderings.[20] Viswanathan took data from a study of nineteen albatrosses that had attached to them a device that took measurements every three seconds and recorded the number of fifteen-second intervals in each hour that the animal was wet for nine seconds or more.[21] Viswanathan was interested to find out how long the albatrosses spent feeding and how long in the air between snacks. He found that the flight times between feeding spots fitted a power law distribution. He speculates that this is the result of the fractal distribution of plankton on the surface of the ocean. Albatrosses feed on the squid and cuttlefish that prey on plankton. The fractally distributed populations of plankton help drive the behaviours of many other species.

It is not just albatrosses that follow fractal forage patterns. The army ants of Southern America and African driver ants are known to do the same. Army ants (*Eciton burchelli*) live on the hoof. Their nests are only built overnight and are constructed out of the living bodies of the ants in the colony. Every morning the foraging ants form raiding columns and fan out to find food and bring it back to the nest to feed the queen and her offspring. The tip of the raiding parties can advance at up to twenty metres per hour. Entomologist Nigel Franks of the University of Bristol has found that this fractal form is the most efficient organization for this rapacious species. Any more spread out and it would be unable to return some of the larger booty to the next spot.[22] Just as in the lungs a fractal framework is the best delivery mechanism.

One of the most striking examples of Universality-mediated behaviour can be seen in the behaviour of the fireflies that gather in the trees alongside tidal rivers in Malaysia, Thailand and New Guinea. Steven Strogatz from the Massachusetts Institute of Technology has spent a lot of time studying these grand constellations of fireflies. During the mating season thousands of male fireflies gather to try and attract females that are flying around nearby. When there are only a few male fireflies around, their flashing is highly uncoordinated. As the night goes on and more fireflies gather the flickering lights become synchronized. 'Eventually whole trees pulsate in a silent, hypnotic concert that continues for hours.'[23] The clouds of magnetism that ripple up and down a ferromagnet at the critical point are made real here.

Group Dynamics

Swarms, flocks, hives, colonies, pods, whoops and flanges. Call them what you will but many organisms live in large social groups. Now it is starting to look like many of them are organized by the principle of Universality.

In ant colonies it seems to help organize behaviour and drive adaptability. An ant by itself is staggeringly stupid and will rush around randomly, accomplishing very little. Put a few more ants together and you get a lot more random activity and no more progress. These small groups of ants find it impossible to learn. Given a forked path to walk down to find food they will always randomly choose one branch, even if the food is always put at the end of the left fork. However, when you put enough ants together you get far more than you might expect – highly adaptive behaviours for instance. Once the number of ants in a colony passes a particular point there is a significant leap in their ability to learn, survive and prosper. Randomness is replaced with purpose.

Experiments have been done with ant species of the genus *Leptothorax* (Latin for 'thin-waisted'), which have tiny nests that can be studied easily in the laboratory. The nests of these tiny ants can fit inside a hollow acorn, but in the lab they are often happy to build one between two microscope slides.[24] Typically there are only a hundred or so individuals in a colony of these ants. Videos of them have revealed that once a critical number of ants is present there is a huge leap in the ability of the colony to support itself. Below this critical amount the ants are running hither and thither and nothing really gets done. With the right number the colony becomes adept at exploiting its surroundings.

Colonies can get organized into even larger groups. Some become supercolonies. Ant metropolises are made up of many colonies in the same way that a city starts as a few towns which gradually coalesece. One of the biggest of these supercolonies exists on the coast at Ishikari Bay in Japan. When first discovered during the 1970s the supercolony was estimated to be home to an estimated 407 million red wood ants. The giant network of nests was thought to be made up of 45,000 smaller colonies and ruled over by over 1 million queens. The colony stretched for twelve miles, bigger than many human conurbations. It might have lasted longer than many cities too.

Professor Seigo Higashi from Hokkaido University believes that it was begun about a 1,000 years ago. He has spent twenty years studying the colony and still does not know why the ants in all these interlinked colonies tolerate each other. They share tunnels and are happy for workers from other colonies to wander in and out of the nests. Now unfortunately the colony is under threat because the expansion of Ishikari Bay is building on top of the colony and preventing the ants communicating. Professor Higashi thinks it is doomed.

The study of the dynamics of groups, colonies and supercolonies has parallels with the work that was done on cellular automata and phase transitions. Chris Langton and others have found that at certain points a cellular automaton shows a vastly improved ability to carry out computations. Brian Goodwin and Octavio Miramontes from the Open University in Britain and Ricard Solé at the University of Catalonia have simulated ant colonies on cellular automata and found the same striking dynamics. Above a critical concentration of ants the dynamics stopped being random and became rhythmic.

This fractal organization is not limited to ants and other highly social insects. In other animals it reveals why they form into groups of particular sizes. Some of these animal groups have been huge. The American buffalo is a case in point. Before it was almost hunted to extinction it was estimated that there were as many as 'sixty to seventy million buffalo roaming the plains between Texas and Saskatchewan'.[25] Contemporary hunters had little feeling for the beasts because the numbers were so vast.

Buffalo hunter William Clark Kennerley – nephew of William Clark, who with Meriwether Lewis had first explored the Far West in the early 1800s – tells the story of coming across a herd that must have numbered in the millions. There were so many of the beasts that it took them two days to pass by the hunting party, who only survived death by trampling by shouting and gesticulating to warn off the beasts and lighting huge fires at night to keep them away. He says: 'I must say that a great many more buffalo were destroyed than was necessary to supply our larder, but what man could resist the temptation when the whole earth, it seemed, was a surging, tumbling, waving mass of these animals?'[26] Few people could resist the temptation, so that by the end of 1882 there were less than a thousand of the animals left alive.

When large groups gather self-similarity is usually seen. Fractal patterns have been found in, for example, the distribution of school sizes for tropical tuna fish and sardinellas. The size of an animal group is partly determined by the ability of the environment to support it. Fertile plains such as the American Midwest were the perfect place for buffalo, and their herds grew to enormous sizes. Herds, however, are not homogenous masses of animals – there is organization here too. Again a fractal distribution of buffalo across a plain or a school of tuna in the open ocean might represent a reaction to the underlying ecology, but it might also, like an ant nest, represent the best organization if information about predators is to be passed around quickly. Any larger and there would be too much confusion; any smaller and the message would not be passed on at all. With a self-similar organization information can be passed swiftly because the whole herd can react swiftly. If we did but look we would find that large herds are in reality made up of groups of roughly the same size clumped together. The loose gathering of a few groups forms a supergroup and clumps of supergroups form a huge herd.[27]

Emerging Evolution

There is limited evidence that populations do rise and fall according to a power law. A 1998 study of American breeding bird populations revealed a power law that held over six orders of magnitude.[28] To arrive at this conclusion Timothy Keitt from the Santa Fe Institute used data gathered over a thirty-one-year period about 600 different species of birds. The results show that larger populations can survive short-term catastrophes and the numbers of birds recover more quickly the larger the population. Relatively rare birds suffered disproportionately in the severe winter of 1975–76. The fluctuating populations show the interplay of local interactions between different species of birds. Beyond individual populations are dynamics that determine who colonizes a spot and who moves on. The highly coupled nature of these interactions reveals that the system resembles a critical point. A small change, such as a large population of one species choosing a new nesting ground, can have consequences throughout the population of birds.

If some animals organize themselves into herds and schools according to the principle of Universality, then evidence from the fossil may reveal that populations of organisms have risen and fallen to the same beat. Unfortunately it is not as simple as that. To begin with the fossil record is not complete. It is estimated that since life on earth began four billion species have walked, crawled, swum or otherwise existed upon it. Each species has hung around for about 10 million years before becoming extinct. At any one time there are estimated to be between 30 and 50 million species in existence. Of those perhaps 2 million have been studied, classified and named. Our knowledge of organisms that have become extinct is nowhere near as good as this. Millions, perhaps billions, of creatures have lived and perished without leaving any trace of their passing. Our knowledge of the lives of dinosaurs is based on less than 300 species, yet these terrible lizards had the run of the planet for over 250 million years. We're obviously missing something.

Imagine if you were asked to work out which sentences the following letters came from: w, t, n, a, t, o, w, s, l, i. With no clues about where these letters fall in the missing words, or how many words you are supposed to be reconstructing, it is possible to make almost any sense you want out of it. You have no way of knowing whether the string comes from:

> The workers have nothing to lose in this revolution but their chains. They have a world to gain. Workers of the world, unite!

or something more poetic such as:

> I wandered lonely as a cloud
> That floats on high o'er vales and hills,
> When all at once I saw a crowd,
> A host of golden daffodils;

Archaeologists are in a similar position with the fossil record. They do not know what is missing, nor how extensive were the species that they do have the fossilized remains of, they have no sense of how big the gaps are in the record and how incomplete our knowledge is. It might also be the case that most of the fossils we find are examples of the unsuccessful species. The ones that persisted may have left fewer traces but in their day may have been the real survivors.

Having said this the fossil record is complete enough to let us draw some conclusions from it. We have to distinguish between what is complete and what is adequate. There is no doubt that the fossil record is incomplete and many fossils have been destroyed by geological activity over hundreds of millions of years. Yet despite the gaps there seems to be enough evidence to draw some conclusions. Unfortunately there is heated debate about which conclusions we can draw.

Charles Darwin was convinced that evolution was a gradual process and that adaptations slowly accreted to give some organisms a selective advantage. This gradualist approach has started to be questioned as evidence of huge fluctuations in the numbers of fossils found have come to light. If Darwin was right then the numbers of fossils found should stay relatively constant as natural selection gently weeded out the unfit organisms. Instead we find there are bursts of births and extinctions punctuating the evolutionary record.

The greatest proponents of this theory of punctuated equilibria are Harvard University palaeontologist Stephen Jay Gould and Niles Eldridge of the American Museum of Natural History. The theory gained credibility in the 1980s by the work of David Raup and John

Sepkowski of the University of Chicago, who, during their work on thousands of fossil genera, found that extinctions tended to happen in bursts. As they see it, most of the time life follows its tranquil course with few upsets to alter the way that organisms live their lives. Then, abruptly, there is a burst of activity when new species emerge and old ones die in large numbers. Under this view 'evolution is boredom mitigated by panic'.[29]

One of the greatest punctuations took place 570 million years ago and is known as the Cambrian explosion. At this point multicellular organisms first appeared, rapidly colonized every available ecological niche, and then got down to some serious evolving. There was another great punctuation at the end of the Permian period, when over 90 per cent of species were wiped out. There is a heated debate over how to interpret this information. In fact almost every flavour of complexity theory has its own ideas about what it means.

Per Bak and other believers in self-organizing criticality say that this pattern of catastrophe followed by periods of tranquility is evidence that SOC is at work. As experiments with sand piles have shown, catastrophes, in the form of avalanches, do not need an external cause. The simple fact that the system is open and energy is being added all the time causes the sweeping changes. The claim that SOC is at work has been bolstered by work with computer models and cellular automata such as Tierra.

Unfortunately there is ample evidence that many of these extinctions had external causes. The late Permian event that left only 10 per cent of all species alive was probably caused by the sea level dropping,[30] and it is now widely accepted that the reign of the dinosaurs came to an end thanks to a huge meteor impact.[31] The huge crater in the Yucatan peninsula is taken to be the site of this impact. While there is argument about the cause of the extinctions almost all palaeontologists accept that they were caused by environmental change.

The claim by proponents of self-organizing criticality that these events could have been caused by basic dynamics of the system is undermined by the fact that they were caused by other external events. Research using other models of extinction imply that there is more than one way to produce the result seen in the fossil record. Simple environmental stress and competition fits the data just as well.[32]

However, if SOC doesn't seem to apply here, then Universality does. The basic theory does not go as far as predicting what will happen, but it does allow insights into how species interact and how the success of one organism puts stress on others. One of the key features of a system at the critical point is that change can happen and spread rapidly. In the case of ecologies and evolution this might mean that one species can easily displace another. Local successes can have long-range consequences: they can echo down the centuries.

Chapter Five

The Business of Complexity

Men, it has been well said, think in herds; it will be seen that they go mad in herds, while they only recover their senses slowly, and one by one.

Extraordinary Popular Delusions and the Madness of Crowds, Charles Mackay

In the long run we are all dead. Economists set themselves too easy, too useless a task if in tempestuous seasons they can only tell us that when the storm is long past the ocean will be flat again.

A Tract on Monetary Reform, John Maynard Keynes

Shopping Science

There is one supermarket that groceries giant Sainsbury's owns that in many respects is utterly unlike any of the other 341 stores that carry its name. It is a store packed with shoppers but none of them is saying a word. The shoppers that know each other exchange greetings using no sounds and whole conversations pass without a word being exchanged. The shoppers prowl the aisles noiselessly, carefully scrutinizing shelves and shopping lists to ensure they get everything they need. Some drag strangely silent children around with them and everyone takes care not to bump fellow shoppers with their trolley. When a trolley jam blocks an aisle everyone queues without complaining while the clog clears. In other respects these silent shoppers are eerily human. They linger in the wines and spirits department, taking time to select the right vintage, but they scoot through the dairy produce picking up milk and cheese as they pass by. They are fiercely loyal to some products and capricious about others. Behind the scenes of this silent shop in unseen warehouses invisible lorries draw up, unload ephemeral stock and leave again. As the food is taken off the shelves and bought by the shoppers, restocking is done by ghostly workers that no one can see.

This supermarket is the only shop these consumers visit. They only come alive when they enter its doors and have no interest in anything else. They have no jobs, no hobbies and never argue with their spouses. All they do is shop. The checkout is the limit of their lives. They are utterly loyal to this store and never shop anywhere else. It is the centre of their world and the sole reason for their existence.

These shoppers never eat, sleep, drink or exercise, yet they return to the supermarket at least a couple of days a week to top up their larders. Once a week they do a big shop and buy many of the goods they bought last time and never used and never ate. As the seasons change the goods that they pile into their trolleys alter. As summer comes on salads replace soups and barbecue briquettes are in demand. Occasionally someone buys a lawn chair.

Although these shoppers have neither faces nor limbs and seem to float rather than walk, they are just like any other bunch of consumers. Today is a hot day and as the temperature passes a certain point some hidden switch in the heads of the shoppers seems to have been flicked and suddenly there is a run on ice cream. The freezer compartments quickly get cleared out and shoppers become agitated because they cannot find exactly what they want. A few grudgingly settle for ice cream substitutes such as lemon sorbet or strawberry cheesecake, but they are not happy about having to make such a choice. No more ice cream is due to be delivered today and many shoppers are going home dissatisfied. With 11 million trips to Sainsbury's stores each week, that could add up to a lot of disappointment.

It should be obvious by now that this is not a description of a real supermarket. In fact it describes a computer simulation of a supermarket that has been created by researchers at Sainsbury's Innovations Centre in London. Although it appears to be false in all the important respects the simulation is peculiarly authentic. It is modelled on one of Sainsbury's North London stores (the company will not say which) and its virtual shelves are laid out in the same way as those in the real store. The fake food stacked on the shelves also mirrors the layout of the real store. The supermarket is considering modelling almost all its stores and its attendant shoppers on computer.

It needs to go into this much detail because the layout of the supermarket can have a profound effect on how successful a store it is. The inside of a supermarket is one of the most tightly engineered environments on earth. Almost everything about the layout of a supermarket is designed to make people spend more. Goods that the supermarket wants to push are put at eye level so that shoppers see them first. The air-vents from in-store bakeries send their warm welcoming aroma around the shop to make people feel at ease and peckish, and hungry people buy more food. Some aisles in supermarkets are kept as choke points so that people are forced to linger and to look at the shelves around them. This way people may buy food they would otherwise never search for.

With a realistic layout and programmable people Sainsbury's can try far more combinations of shelves than it ever could in an actual store.

Shopping for Answers

Although the programmable people populating the SimStore, as the simulated store is known, do not talk and float rather than walk, in other respects they are as realistic as possible. The shopping lists that they enter the store with are as close as possible to lists real people shop from. Sainsbury's has drawn up the lists using information gleaned from the loyalty cards the company gives out.

Anyone signing up and using the card amasses points for every pound they spend. In return they get money-off vouchers or coupons and a host of offers denied to customers without a loyalty card. But Sainsbury's receives far more in return. It gets reams of detailed information about spending habits, what people buy at all times of the day and, even better, it can relate this to post codes, addresses and other demographic information.

Just as important as what people are buying is when they buy it, how often they visit the store, and what combinations of goods they buy at the same time. By using this data Sainsbury's can make all kinds of predictions about shopping habits to try and ensure that the shelves always hold what people want and aren't empty when they should be full. 'Using the data we can predict what kind of alcohol is likely to be bought by pet-owning parents who shop on a Wednesday evening and Saturday morning,' says Mark Venables, director of the Innovation Centre. Information like that is very valuable to Sainsbury's and helps it tune deliveries to meet demand. Empty shelves mean unhappy shoppers and lower profits. The collections of data have already revealed some curious buying habits. For instance shoppers who regularly buy four-packs of Coke are happy to buy eight-packs when there is no more of their usual size, but they would never buy Pepsi.

However, the simulation is likely to be of most use in revealing the pivot points of human decision-making. Like many other situations, when you put a herd of people together the pattern of decisions that emerge from the crowd shares some of the characteristics of a system at a critical point.

Many of the decisions that shoppers make are easy

to understand. There are a lot of foods, such as bread, milk, and the like, that people buy every week. For many other foods, such as seasonal items like ice cream or fashionable foods such as pesto, it is hard to predict when tastes will change. When they do change, they do so rapidly and often for no immediately obvious reason. It is often hard to point to just what started the panic off. Like an avalanche spreading down the slope of a sand pile or a herd stampeding, suddenly a society is on the move.

Venables and Sainsbury's hope that the SimStore will give them an insight into some of the dynamics behind those changes of taste. Temperature is obviously one important variable. As many supermarkets have found to their cost during a run of good weather, ice cream sales will steadily trundle along until the mercury climbs a degree or two past an unmarked but widely shared point, at which point everyone wants to buy the stuff.

Like all models of systems at their critical point it is not the properties of the individual elements that are important, merely the dynamics driving the structure of their decisions. Silent shoppers may never say a word but they act just like you and me.

Here Today, Gone Today

Sainsbury's is not the only company turning to simulations to get a better understanding of the dynamics of shopping. Giant technology and management consultancy PricewaterhouseCoopers is also experimenting with models of buyer behaviour but on a much grander scale than those that Sainsbury's is building. In the splendidly titled Emergent Solutions Group at PWC, Winn Farrell and his team are attempting to model and understand how hits happen, be it a hit movie, a hit track or a hot book.[1] They want to see if the forces at work behind an insanely popular record can be understood, teased out and manipulated. As Farrell says, his group is trying to create 'a flight-simulator for shopping'.

No one really understands what it was about Dava Sobel's book *Longitude* that turned it into such a huge hit. The latest estimates are that *Longitude* sold around 1 million copies in hardback alone; eventually paperback sales will exceed that. Nor does anyone know what catapulted *The Full Monty* into its huge success. *The Full Monty* was the most popular film of 1997 and was the most successful UK film ever until the 1998 release of *Notting Hill*, starring Hugh Grant and Julia Roberts.

The relationship between the content of a book, film, record or computer game and its subsequent success is poorly understood. Fads, fashions and trends are decidedly non-linear. If there was a straightforward or linear relationship between the amount of money and time a company spends promoting a toy, book, album, movie or game and its success then there would be no surprises. Like preprogrammed sheep we would react to the marketing and buy the things that we are told are good. Except of course that we don't. Much to the chagrin of marketeers around the world the public are usually impossible to second-guess. Usually.

Farrell and his fellow researchers at PWC think they

are close to plotting just what it is that makes a toy, record, book, or whatever into a palpable hit. They are starting to find out the key inflection points where a tiny change causes an enormous response. It might be just a case of ensuring a single is played fifty-two times rather than just twenty-six. The result could be sales of 20,000 copies instead of just a few thousand.

Already Farrell claims the group has created a simulation that can predict first-week box office receipts to within 80 per cent of the actual figures. Instead of shopping lists this simulation is fed with market data about where the film is getting mentioned, who is talking about it and how often the news about its release is being repeated. In effect it recreates the conversations, TV viewing and radio listening habits of around 150,000 people. As with Sainsbury's SimStore the programmable people are as realistic as possible. They are programmed with different ages, incomes, homes, sexes and buying habits. Some are fickle and change their tastes quickly; others are just huge fans and buy everything one artist produces; still others are only convinced that something is good if most of their friends do. The data that feeds Farrell's flock is data culled from market research that makes them as cussed, irrational, up-to-date or out-of-fashion as the general public.

Farrell and his colleagues have found, just like the physicists studying hot ferromagnets, that small changes can have enormous and far-reaching effects. The flipping of one tiny domain in a magnet can trigger a cascade of reactions throughout the entire magnet. The effects of long-range correlations can be felt far from the origin of that tiny change. Similarly, if the right person likes a film or the right group of people go and see it, then a work that is relatively unknown can go stellar. Like the elements of a system displaying universal properties everything is connected. It does not take many steps to span the world.

The important elements seem to be the networks of relationships between people. 'What is very important are interconnected chains of events – I know you and you know him. That allows the social networks to become alive,' says Farrell. 'These are very important around software and movies.' These 'network events' can boost the success of a film, book or record.

A perfect example of this is the success of the band Hootie and the Blowfish,² who in early 1994 were relatively unknown. This bar band from South Carolina often found themselves supporting other bizarrely named groups, such as Toad the Wet Sprocket. Yet by December 1994 they were huge. The band's debut album *Cracked Rear View* roared to the top of the US charts. They made the cover of *Rolling Stone*, appeared no less than three times on the David Letterman show, and by the end of 1995 their album had sold around 13 million copies. They became one of the fastest selling bands in the world. Even if you didn't like their music you had to admit that, by anyone's standards, they were superstars.

Farrell believes that Hootie and the Blowfish benefited from these network events, which conspired to create a buzz about the band. People who went to see them in concert told their friends that they were great live and persuaded them to buy the CD. Liking what they heard these folks told others and slowly the momentum gathered. It might be the case that many people who liked the band thought that they were going to be big and wanted to be one of the folks who picked a winner. Whatever the reason the effect was the same. Farrell says this interest created an 'information contagion' which propelled the debut album to the top of the charts.

In some respects these information dynamics are similar to those that are at work when an epidemic is sweeping through a population or a forest fire is raging. The dynamics behind spreading diseases and fires undoubtedly have a lot to do with Universality, but it is harder to establish that this is the case for record sales because of the difficulty of following patterns of taste. You can't spot sudden fans as easily as you can see that a tree has burst into flames or people are being admitted to hospital. However, what evidence does exist is persuasive. These dynamics among people seem to be at work at all scales too. Not only do they work within groups of friends but they seem to work when people go shopping too. Farrell says that studies of buying trends have revealed bursts of people buying the same thing, be it a book, record or toy. It is likely that some subtle communication is going on. Perhaps seeing someone carry a record reminds a shopper what they were looking for.

By plugging information about how often Hootie records were played on radio stations to who was buying the CDs and where, the simulation was able to reveal how the waves of demand grew, merged and became unstoppable. 'Hootie enjoyed the physics of the hit,' says Farrell.[3]

Hope not Hype

The simulation clearly captures these non-linear physics. Farrell and his team have used it to study two other 1995 hits, by Rod Stewart and Seal, and found they could recreate sales histories to within a few per cent of actual sales.[4]

It is interesting to note that once the success of Hootie and the Blowfish got past a certain point people were no longer buying the records for the music. They were buying the CD because to be without it was to label oneself uncool. This suggests that sometimes the increasing success of a record or film has nothing to do with how good something is but is due to the buzz surrounding it.[5] People buy it because everyone else is. Critical reactions no longer matter. There is a strong resemblance between this kind of behaviour and that found at the critical point where organization, structure and interactions matter more than the properties of the elements in a system.

This might go some way to explaining why many bands enjoy great success with their first album but struggle to do the same with the second. The 'difficult second album' might be hard to sell because, if truth be told, the success of the first didn't have that much to do with the quality of their music. It had more to do with the fact that people wanted to be seen listening to it than it did the quality of the sound. Sales of second albums may not do as well because people actually do judge it on the merits of the music. The people who bought the first album because everyone else did may realize when they hear tracks from the second album that they don't really like the music after all.

It might seem that these simulations make people look like automata, robots who react when the right buttons are pressed. While it is true that the simulations are helping companies and marketeers boost sales they are not infallible. In fact the conclusions that can be drawn from this work are quite hopeful. They imply that people are very independent and hard to second-guess. Farrell says that the simulations cannot guarantee a hit. In fact the best they can do is show people what *not* to do. They show the pitfalls to be avoided.

PWC has been talking with one US TV station (for reasons of commercial confidentiality it will not say which) and was contemplating creating a TV show that was tuned by all the market research data used in the simulations to be as hip as possible. In this way, by tuning it into popular currents, they hoped to ensure that it always garnered a huge audience. Of course, if people found out that they were being manipulated, the dynamic that kept it popular would be destroyed and the show would tank and never be seen again.

The most positive message to come out of this research is that it is impossible to hype just anything into success. While the mechanisms that promote success can be used to create demand for a film or book, people are hard to fool. They are remarkably resistant to manipulation because they are so capricious.

Farrell says that to create a hit you have to have that extra something. 'With a hot toy or a book you need content that surprises and delights,' says Farrell. You can't promote any old rag or rock into the must-have item. The hardest part of having a hit record or writing a bestseller is as hard as it ever was. It remains frustratingly difficult to predict just what people will find surprising or delightful. The second most important quality a hit requires is the ability to deliver on promises. If the marketing campaign creates interest that is never fulfilled then this will show in the book, film or toy's subsequent success. *Star Wars: The Phantom Menace* was the perfect example of this.

But, says Farrell, if what you create has those qualities, if you have a well-written story, a catchy tune or compelling plot, then the simulation shows that it may not take a great deal of effort to reap a huge reward. Just ask Hootie and the Blowfish.

Life in the Fast Lane

The reason that Mark Venables at Sainsbury's and Winn Farrell at PWC build these simulations is because there is no other way to study the problems that they want to solve. If there are too many correlations and interactions at work in a ferromagnet kept at its critical point to explain with mathematics, then this is even more true of people and populations.

People are not bipolar like tiny grains of magnetism: they have more than two opinions and don't swing wildly between extremes. Our opinions, beliefs, likes and dislikes span many spectra and shades of agreement. There are many subjects that we care passionately about and others that, frankly, we don't give a damn for. In-between lies most of our mental life and it contains our thoughts on issues that we have half-formed semi-debated opinions about. Some beliefs come from our upbringing; others are shaped by experience and a few we were born with. Basically it is a right old mess.

Mapping changing tastes is a far trickier task than measuring earthquake tremors. With an earthquake you have the Richter scale to help you quantify how big and bad an earth tremor was. Even if a quake was the end result of a series of minor quakes you have its behaviour captured in an easy-to-express form. The same is true of many other systems that display universal properties, such as heartbeat patterns, hot bar magnets, and so on. The same cannot be said for the tastes of the music-buying or movie-going public. Perhaps there is more of psychology to it than physics. With these kind of systems it is much more difficult to establish that the dynamics at work are the same as those that are found in classically universal systems.

What can be said is that the forces determining hits and misses are of the same sort as those that drive stock market movements (of which more later). There are many thousands, if not millions, of elements involved – stock traders in one instance and consumers in the other. The system constantly has new energy driven into it. In the stock market this takes the form of changes in fiscal policy and which stocks other brokerage houses are buying and selling. With records the tuning comes from marketing campaigns, the number of times people hear the record on the radio or TV, and the preferences of our peers.

With stocks and shares there is no doubt that we are seeing a system displaying Universality. With such things as record sales, hit movies, hot toys and best-selling books the picture is suggestive rather than conclusive. For this reason there is no way to understand people except by building your own and letting them get on with it. This is exactly what Venables and Farrell are doing.

They are lucky that when modelling such complex systems you do not have to reflect every last property of your subject. So programmable people need information about how often they have heard a record but they do not need ears. They need to know that the temperature has risen and that they like ice-cream. They don't need a central nervous system and a stomach. All they have to do is get a few things right and the rest follows.

Simulations that produce realistic results are also valuable because they can speed up time. It is far easier, and cheaper, to get a synthetic population to react to a marketing campaign than it is to do it in the real world and measure the results. The plastic people don't mind being interrogated, they don't worry about their privacy being compromised, and they will happily share every last detail about their lives.

While they are talkative these plastic people are not

predictable. As we have seen in complex systems tiny changes can have very long-lasting effects. Subtle differences in the set-up of every run of a simulation mean that it is impossible to predict that what is successful once will always produce the same result. The simulations reveal the importance of contingencies such as luck. Writing a hit record is not just about clever lyrics: the time has to be right for it too. It is a scary thought but perhaps without the Bay City Rollers punk would never have happened. The simulators show that nothing is certain and there are far too many unknowns to make it an exact science. Probably the best they will be able to do is help companies maximize the impact of a book, toy or record once they have seen it is taking off.

To explore the spectrum of all possible outcomes you have to let the simulator run, and run, and run. The same initial conditions may not lead to the same outcome. The tuning of the system can be tinkered with to see how it can be driven to where you want it to be, to find out just what the physics of the hit entails. The pivot points of the whole system, and the fluid-like map of people's tastes, become more obvious. Once you know where these points are it becomes easier to ensure a book or record becomes a hit. They let you know when the iron is hot so you can strike.

These simulations are also being used because only now is it possible to build them. Computer power has got cheap enough to make it viable. Five years ago it would have taken supercomputers to crank through the interactions of thousands of agents; now a large desktop computer can do it. Some of the simulations that are being built are huge. One such is a digital replica of the New Mexico town of Albuquerque. The model, which goes by the name of Albuquerquia was created by Los Alamos National Laboratory researcher Chris Barrett to study traffic problems in this dusty southwestern town. The simulation is astonishingly detailed. Every one of the 200,000 households in Albuquerque are represented, as are its 400,000 daily travellers and 30,000 road segments. The traffic in this town has been extensively studied and virtual vehicles drive along the computerized carriageways in the same numbers and travel to the same destinations as their real counterparts. The model can be used to see what happens if roads are closed for resurfacing or to investigate the effect of building new bridges or creating a one-way system. Individual vehicles can be watched as they make their way through the town or traffic jam patterns can be captured. All the detail and data is there; you just have to decide how you want to slice it.

Simulations have been used to study smaller-scale problems too. One such is the El Farol problem first formulated and studied by Brian Arthur from the Santa Fe Institute. The El Farol is a bar on Canyon Street in Santa Fe. In the early 1990s Irish music was played every Thursday night at this bar. Irish-born Arthur liked to go along to listen to the music but often El Farol was too noisy and smoky for him to properly enjoy it. In Spanish a *farol* is a bluff or steep bank eroded by a river.

Every Thursday presented him with the problem of whether he should go along. Unlike many others Arthur saw this as a problem for simulation. He took an imaginary population of 100 residents of Santa Fe who, like him, occasionally go along to the bar but don't like it to be too noisy. In deciding whether to go along to the bar these surrogate people used memories of what it was like last time it was there, who was playing, who else they knew was going along, as well as many other criteria such as if the home team won this week. There is no single chain of reasoning that will get it right every time; you have to sift the evidence anew almost every time.

In the long run almost the same amount of people may turn up to the bar every week but on any individual Thursday it will be impossible to predict how many there will be. The people that go along will do so for such different reasons that it becomes impossible to capture such behaviour in any mathematical equation. The only way to study it is through simulation.

Such a model has wide applicability in economics that often tries to work out how people reach buying decisions. Economics, erroneously, takes an average attendance figure to mean that people are using one tactic and sticking to it. In reality numbers fluctuate widely and tactics change quickly. Typically they use informal networks of information to predict what will happen and then decide on that basis. If their prediction turns out to be accurate, they use that method again; if not, they change their tactics.

In early 1998 the El Farol problem was effectively solved, but not because of a breakthrough in theory or a particularly detailed simulation. The bar burned down. Actually this didn't solve the El Farol problem as much as replaced it with the Second Street Brewery problem or the Dave's Place problem.

Lastly, simulations are useful tools for managers and bosses unfamiliar with more mathematical ways of modelling a business or a marketing campaign. All they need to do to grasp what works is watch the simulation. It would be a foolish businessman who used simulations like these to make every decision, but he would be a bigger fool if he ignored all the insights they reveal.

Fractal Finances

It might be stretching the truth to claim that the dynamics of hit records, books and toys are a study in Universality. With no measures of magnitude to grasp, it is hard to claim that a self-similar pattern is at work. The same cannot be said of stock markets and the rise and fall of other prices. Even better there is a wealth of data to hand that can be analysed to see what patterns are hiding within.

In fact Universality and the methods that physicists have developed to explain its dynamics are proving so useful that it has given rise to an entirely new field: *econophysics*, a term coined by Gene Stanley. As the name implies this treats the economy and the fluctuations of economic indicators as if they were problems for statistical mechanics – a gas, or a hot magnet perhaps. Universality is the reason why it makes sense to do this. As we have seen, systems at critical points have exactly the same dynamics and properties. The techniques used to describe and explain one can be used for all of them. While the name has only been coined in the last year or two, the evidence that the financial world is a critical system has been building for over a century.

The first scientist to discover power laws in finance was Italian social economist Vilfredo Pareto. While working in Switzerland he noticed that a power law could be seen in the number of people in a stable economy whose income exceeds a particular large value. He found that the result was quite general and held across societies as different as England, Ireland, Germany, Peru and many Italian cities.

However, it is Benoit Mandelbrot who has done most of the work on establishing the existence of self-similar or fractal patterns in financial indices. In fact one of the many reasons Mandelbrot became convinced that reality was fractal was because of his work on economic indicators. The ubiquity of fractals was brought home forcibly in 1960, when Mandelbrot took up an invitation by Harvard economics professor Hendrik Houthakker to talk about his work.

At the time Mandelbrot was studying the range and distribution of incomes in an economy – how many rich people there are and how many poor. Houthakker was struggling to make sense of the fluctuations in cotton prices and wanted to see if the young mathematician could shed any light on his work. When Mandelbrot turned up at Houthakker's office he was astounded to see his findings already written on the blackboard. In fact the figures he was seeing were no such thing.

Chalked up were eight years of changes in cotton prices, not figures representing the distribution of incomes. At first glance, however, they looked the same. They were self-similar. Houthakker was studying these figures because they represented something of a puzzle. Usually variations come in a handy package known as a bell curve, but Houthakker could not fit the changes in cotton prices to any kind of bell curve.[6]

Mandelbrot noticed immediately the similarity with the data he was working on and once the talk was over rushed back to the IBM Yorktown lab and ordered records of cotton prices dating back to 1900 from the Department of Agriculture in Washington. He set about analysing the results and was astounded at what he found. This fractal pattern of price changes held true over sixty years of variation that took in two world wars and a depression. Further work by Mandelbrot has shown that this power law holds across nearly a century of data.[7]

Since then self-similar or fractal patterns have been found in an enormous number of financial indices. Power laws have been found in the prices of nineteenth-century and modern securities (stocks) and twentieth-century interest rates. It turns out that they are everywhere, ubiquitous, universal.

Model Money

Initially the results of Mandelbrot's work were a couple of papers published in 1963 to little effect.[8] It is easy to see why. When these papers were published the pioneering work by Leo Kadanoff, Michael Fisher and Kenneth Wilson on phase transitions had not been done. There were no methods that could explain the dynamics of a ferromagnet at its critical point, let alone something as fiendishly complex as a stock market.

Even when Wilson's pioneering work had been done it was only applied to problems in statistical mechanics largely because at that time physicists rarely looked beyond the boundaries of their own discipline. They had a hard enough time explaining the dynamics of systems they could control and had no desire to make life hard for themselves by tackling a subject as formidable as economics.

It was the explosion of interest in chaos theory that gave physicists and mathematicians the courage to tackle stock markets and the movements of prices. One of the key insights of chaos theory is that unpredictability can be the result of orderly or regular input. Heat a liquid from below and convection cells will form as the liquid rises, cools and then falls. Keep heating and the cells will start to wobble and then shatter into a turbulent maelstrom. The heat drives it to chaos. The same was thought to be happening with stock markets. Orderly input – the regular buying and selling of stocks – was believed to be producing essentially random movements in market indices such as the FTSE 100.

However, a couple of decades of work has failed to prove that chaos is at work in financial markets. Instead financial markets have been revealed as canonical complex systems and as such understandable (if not explainable or predictable) using Universality. Financial markets share many of the characteristics of the systems that Universality has helped to tackle. They are open systems but instead of energy being fed into them traders drive the system by buying and selling stocks. They are made up of thousands of essentially identical elements: traders. These units interact in a non-linear fashion, i.e. the decisions of one unit can have huge consequences. The units react to previous market movements, providing feedback and keeping the system turning over.

In other critical systems this complex set of inputs produces a simple result: price movements that follow a power law. This is the reverse of chaos, where simple dynamics reveal highly complex results. Here a very complex system produces a very simple result: a power law distribution.

Now that physicists have realized that stock markets can be treated in this way the evidence for those simple results – self-similar patterns in price changes – is being found everywhere. The search for evidence has been made much easier because since the 1980s the movements of stock markets and exchange rates have been monitored using computers.

Currencies are traded twenty-four hours a day and data is gathered about exchange rate movements on a minute by minute basis. In some markets movements are monitored even more closely. The same is true of almost any financial instrument you care to mention – derivatives, swaps, options. It does not matter. The ups and downs of the markets are endlessly logged.

Now there is an embarrassment of data available for

analysis as the volumes of trades has increased enormously. For example, the volume of foreign exchange transactions carried out in 1995 was eighty times what it was in 1973. The growth has been even more dramatic in financial products known as derivatives because their value is derived from underlying assets such as stocks on front line markets such as the FTSE 100 and NASDAQ. In 1996 the total value of financial derivative market contracts issued was $35 trillion dollars. Anyone interested in studying the changes in prices is spoiled for data. There are no worries about having too few orders of magnitude to establish if fractals are present.[9]

Gene Stanley and Rosario Mantegna conducted one of the most exhaustive studies of a financial market, in this case the Standard and Poor's (S&P) 500. This index is calculated by adding together the market capitalizations of 500 companies said to be representative of the US economy. As with any index of financial activity the movement of the S&P 500 index is recorded regularly, sometimes as often as once every fifteen seconds. Stanley and Mantegna analysed data records from a six-year period giving them over 1.5 million points of data. The analysis involved noting the size of each move, either up or down, and then counting how many changes of each size could be found in the data.

Stanley says this can be thought of as creating 1.5 million sticks each one of which corresponds to the size of each rise or fall in the S&P 500 index. The sticks of each length are then counted and their distribution is plotted on a graph. This reveals how many jumps of each size there have been. Most of the changes were small, there were a few medium-sized ones, and big jumps were rare.

Just as Houthakker and Mandelbrot found, the distribution did not fit a bell curve. Instead it was a self-similar pattern known as a Levy distribution. This distribution of events resembles that for earthquakes in which large events follow a power law. Mandelbrot found the same sort of distribution in those cotton prices.[10] The power law held true over more than enough orders of magnitude. At every scale between seconds to minutes to hours to days to years the relationship held true.

As if that was not enough Stanley and some colleagues from Boston University then did another even larger study. This time they took data from the Trades and Quotes Database. This contains records of who is buying and selling stocks and how this affects their prices. The database does not hold records of all the stocks traded in all the exchanges in the US but it does capture all the deals made in the country's three largest stock markets. Specifically it records the deals done and stock market movements on the New York Stock Exchange, the American Stock Exchange and the National Association of Securities Dealers Automated Quotation, or NASDAQ. Between them they constitute a significant chunk of Western economic activity. Shares in many of the largest companies in the world are traded on these three exchanges.

Stanley and his team didn't use all the information from this database. Instead the researchers concentrated on movements in the stocks of the 1000 largest companies over a two-year period from January 1994 to December 1995. This span of time gave them 40 million data points to work with, by far the biggest swathe of data ever analysed to see if power laws could be found within it.

The team used the same technique to analyse this data that was employed in the earlier study. They worked out the size of each stock movement and then counted how many jumps of each size there were. Once the data was plotted on a log–log graph, it was easy to see another power law at work within it.[11]

Economics Overturned

On the face of this evidence it would seem to be a simple step to conclude that Universality is at work in financial markets and economies around the world. Certainly in some important respects financial markets do resemble the experimental systems that have helped to establish Universality as a theory.

Like a ferromagnet held at the critical point a stock exchange is an open system that has a constant supply of energy flowing into it. In the magnet it takes the form of heat but in the stock market it is money and changes in financial policy. The buying and selling of shares constantly gives the market more money to play with and traders a reason to deal. Adjustments to interest rates as well as tax and legislative changes are another important source that helps keep the flow of money across the markets more or less constant and tunes them to a critical pitch.

The units of the stock market – brokers and traders – constantly interact as they buy and sell shares for themselves or their clients. This feedback creates a roiling web of activity in which small changes can have huge consequences. There is no linear easy-to-grasp relationship between the number of deals done and the movement of the index. Markets can plunge if the wrong deal is done at the right time.

In some respects this resembles the 'network effects' drawn out by Winn Farrell and his colleagues at PWC. Just as with a Hootie and the Blowfish album when people started reacting to the buzz around the band rather than the music, so in a stock market brokers and speculators often react to what everyone else is doing rather than decide for themselves. If a stock is soaring or plummeting traders will sometimes join the stampede with no regard to the strength of the company behind the stock. People just want to join the gravy train or sell what they own and get out.

Despite these signs it is perhaps too early to say that mainstream economic models are useless. To begin with economics differs radically from physics in the units used to measure the movements of markets. In physics a lot of time and effort is put into determining the best reference units so that all measurements consistently use the same yardstick. Once the unit of measurement is agreed then readings are taken at regular intervals. In finance the opposite is the case.[12] The scales used to measure movements are usually given in currencies. Thanks to foreign exchange trading the value of these currencies changes almost as often as the markets they are being used as the baseline for. They are also being eroded by inflation and subject to other pressures, depending on whether the economy is growing or shrinking. To make matters worse deals are struck at any and all times rather than at the regular intervals physicists prefer. Finally shares in companies are not bought in fixed units. Traders can buy any or all of the shares held by another broker. They aren't limited to buying a fixed amount every time.

All this makes it very difficult to say with certainty that here is yet another system where Universality applies. The field is too young. The best we can do is say that the evidence is persuasive and wait for more analysis to be done. It might be though that the world of finance is a system permanently caught on the cusp of the critical point.

Even if it is right then it remains to be seen just how much use it is to know that the distribution of stock market movements draws out a self-similar pattern. To begin with power laws are purely statistical. They say nothing about when the next big or small event might happen. No one is going to get rich using power laws to characterize movements in stock markets or any other financial indicator. Though some are trying.

Some of the pioneering work on chaos theory, complex adaptive systems and non-linear physics was done by Doyne Farmer and Norman Packard. The eclectic education that this fearsomely smart pair underwent prepared them for working across the usual strictures, and structures, of scientific thinking. Both made their name studying systems that degenerate into chaos or produce order out of seeming randomness. Initially they applied this theoretical work to a rather frivolous end. They tried to beat the casinos in Las Vegas. They used their knowledge and a cabal of friends who were experts in electronics and physics to design computers that could fit inside a shoe and be operated with the toes.

The pair realized that although most of the games in a casino are weighted towards the house, there is one, roulette, that is less random than the others. Anyone who spends enough time studying the physics of each roulette wheel will be able to find out which numbers it favours and place bets accordingly. The shoe computers were there to record the behaviour of the wheel and give hints about where to put the money. The adventures of this group were chronicled in a book called *The Eudaemonic Pie* by Thomas Bass.[13] It's a good read, if exasperating, because although the group spent years refining their shoe computers and won a little money, they never got close to a big score.

Now Packard and Farmer have moved on to found the Prediction Company, where they are attempting to apply their expertise to another gambling den: Wall Street. The pair believe, and so do the people they have gathered around them, that, although stock markets are inherently unpredictable, they should be able to make short-term predictions or spot windows of order that they can exploit. Thomas Bass has been the chronicler of this tale too in a book called *The Predictors*.

The Prediction Company is developing software that will help stock traders beat the market every now and again. This time around they appear to be much more successful. They have the financial backing of the Swiss Bank Corporation. On several occasions they have made a million dollars during their lunch break.

However, it took them a long time to get to this point. The software tools have been under development for seven years. They are employing genetic algorithms and other adaptive systems to spot the eddies in the market. The first few times they tried the system it failed. It was caught out by events that no one could predict. The Federal Reserve Board announced a surprise change to short-term interest rates and the Prediction Company lost a lot of money very quickly. Every so often the same thing will happen, but for the moment Farmer, Packard and co-workers are making money.

All that power laws do is reveal the bounds of the largest and smallest movements but they will not reveal when the next change of any specific size is due. It does set some very broad limits on the size of changes but that is about all. The best it can do is give you a hint over very short time scales about the direction of movement. This might help the largest brokerage houses, which have huge assets to play with, but the rest of us will have to take our chances.

This means that we are going to have to learn to live with stock market crashes. Sudden sweeping changes are a key characteristic of systems at a critical point. At once such systems are both robust and skittish. The constant addition of energy keeps them ticking over with events of different sizes popping up all the time. The whole is a rippling seething mass. Then for no apparent reason one tiny change causes an enormous shift throughout the system and a stock market plummets. There is nothing you can do about it and no way to predict it: it just happens. Those wild gyrations will happen no matter what you do.

There is little point in trying to dampen the effects of these swings because they are built in. In fact some scientists suspect that they are going to get worse as stock markets become ever more automated. We have seen earlier that if a system is driven too hard then it can collapse into overwhelming chaos. Some fear that this might also happen to any market when governments try to take pre-emptive action.[14]

Even if we cannot predict then we can at least prepare. If we know that stock markets will always follow a pattern of boom and bust then we can try and ensure that we, by which I mean individuals and whole economies, are not caught out. If anyone living in a zone regularly shaken by earth tremors never did anything to prepare or build houses that could survive the smaller shocks we would criticize them for their refusal to face facts. The same should be true of anyone buying and selling shares. Be prudent: don't lose what you cannot afford and regularly reduce your holdings. By converting shares to cash you should be insulated from the ups and downs of the market. It might not be as exciting as risking everything in a single trade in a roller-coaster market but if you don't prepare – just like people who live in areas regularly wrecked by earthquakes – you might be left without a roof over your head.

Econophysics is helping some people save money, but not perhaps as much or for the reasons you might expect. If it is the case that financial markets share many of the characteristics of much simpler physical systems then this opens up new ways of studying them.

Currently most stock traders rely on historical trends and number crunching to try and get one up on the market. They use huge amounts of computer power to analyse trends to try and spot when the next boom or bust is due. They run and rerun scenarios and pick the best-performing strategy. Universality implies that this is a fool's errand, but so far this does not stop the traders trying.

Instead the traders and research houses should adapt or adopt the methods physicists use to study or replicate those simpler physical systems. Hideki Takayasu and his colleagues at the Sony Computer Science Laboratories are leading the way to better simulations using a simple electrical circuit that can mimic yen–dollar fluctuations. Rather than costing hundreds of thousands of dollars this circuit costs only $5 and is just as good at generating fractal sequences of currency changes. The circuit uses naturally occurring electrical noise (the stuff that Shannon and Mandelbrot were so interested in) as seeds to generate a fractal series.

Be Economical with the Truth

If the promise of this early work is borne out and Universality is shown to permeate finance then it holds a sobering message for economists. Specifically it says that their view of the world is wrong. The cherished theories that economists have used to characterize the movements of markets are utterly useless, largely because they abstract away all the details that Universality shows to be so important. Like physicists at the beginning of the twentieth century economists are clinging to an outmoded and old-fashioned view of the world.

To an economist the price of a good, be it a gun or butter, is determined when supply and demand are balanced. There is an assumption that the blind workings of the market will find the price people are prepared to pay, given how rare those goods are, the cost of producing them, and how much they are desired. In this model people are assumed to have a perfect understanding of their wants and needs, as well as all the information they need about the things they want to buy. As the British economist Lionel Robbins put it: 'Economics is the science which studies human behaviour as a relationship between ends and scarce means which have alternative uses.'

In this world prices change smoothly as supply and demand wax and wane. Prices may move up or down but the balance between supply and demand will always be maintained. Small changes, such as a slight hike in interest rates, causes a small change as the economy and the interlocked forces of supply and demand move to compensate. All the jagged edges and surprises of life are smoothed away or compensated for. This model assumes that there are no market crashes and compensates for them. It also assumes that companies compete on an equal basis: there are no monopolies and there is no price protection. It places a touching faith in the magic of an unfettered free market that will ensure everything works out for the best.[15] Except that it doesn't.

As we know people are messy, irrational and stubborn beings. Faced with a series of choices they do not all react the same way, nor are they even consistent themselves. They make different choices from week to week and even from one day to the next. As Sainsbury's is finding out with its programmable people we are very hard to second-guess. Our likes and dislikes seem to move to a rhythm that we know but cannot articulate. People don't react in a rational manner all the time and any theory that doesn't take that into account must be wanting.

Perhaps because the sources of human decision-making are so hard to catch, economists have not really tried. Economists may retort that even if everyone acts irrationally and makes uninformed choices it doesn't matter. What does matter is the actions of the group, and, on the whole, in groups our choices are predictable. This is captured by the *law of large numbers*, which says that if you get enough people together then, like sheep, their movements and choices are easy to predict.

The essence of Universality is its insistence on the importance of details. While the models of the systems being studied can be simple they still have to reflect that sensitivity. Tiny changes must be able to have enormous effects and correlations between the constituent elements must be allowed for. Certainly all the statistics imply that the economist's model of the world is flawed. The prices of shares on stock markets do not change smoothly; there are sharp discontinuities from moment to moment.[16] Rather than deal with these abrupt changes economic models regard them as anomalies and ignore them, saying that they have little or no long-term effect.

Furthermore market crashes are still very much with us. At the time of writing the US has gone through the longest boom in its history. It began in March 1991. Many economists thought that the productivity gains brought about by technology meant that the days of boom and bust were over. It was thought that the cycle had been broken. However, these ideas fell flat in early April 2000, when some of the US markets suffered their sharpest falls ever, and technology stocks in particular took a battering. At the end of 2000 the NASDAQ has lost 50 per cent of its value and a recession is looming. Obviously market crashes are very much alive and well and we still need a way to explain them.

If Universality is at work in finance then perhaps economists should adopt models that reflect this. In many respects the ideas of economists resemble the mean field theory that was found wanting when physicists tried to use it to explain events at critical points. For a long time now physics has been moving on and trying to engage the real world. Current interpretations about the forces at work in systems undergoing phase transitions are based on the work that was done to explain these dynamics.[17] When the theory could not properly explain, physicists changed their theories; economists seem bent on ignoring those parts of reality that do not fit.

If Universality is found to be infecting finance then a radical change in policy might be needed. We should worry less about predicting crashes or trying to mitigate their effects and do more to prepare for them. An attitude of tolerance and acceptance would reap more dividends (literally and figuratively) than a Canute-like attempt to hold back the dynamics underpinning a whole economy.

The Law of Fraud

The question of whether Universality can be found in financial indices remains to be decided. The debate about its presence or applicability is the same as that over fractals and orders of magnitude. At the moment it comes down to a decision about the evidence that you are prepared to accept. Some scientists need more convincing than others. However, there is one financial activity that everyone agrees that there is a law operating at all scales: fraud.

If you want to spot a fraudster, someone who is cooking their books, all you need to do is to go through the sales figures of the company that you suspect and then count how many begin with the number 1. If more than 30 per cent of the sales figures begin with the number 1 then something is up. Generally, the higher the percentage that begin with a one, the greater the likelihood that something fishy is going on.[18] The reason for this is Benford's Law.

The law was discovered by accident. In 1881 the American astronomer Simon Newcomb wrote to the *American Journal of Mathematics* about the books of logarithms used by scientists. He had noticed that the opening pages of these books tended to be much dirtier than the last ones. This didn't mean that scientists were being lazy and avoided doing calculations that required them to turn over lots of pages. Newcomb realized that the scientists were doing more calculations with more numbers starting with 1 than, say, 8 or 9. This was why the pages at the front of the log books got dirtier quicker.

For nearly sixty years Newcomb's observation went unremarked. Then in 1938 a physicist at the General Electric Company called Frank Benford rediscovered it, but unlike Newcomb before him he did not stop with log books. Benford scoured newspapers for lists of numbers to see if they followed the same law. Perhaps unsurprisingly they did. He collected 20,000 numbers covering everything from the drainage areas of rivers to baseball statistics, numbers appearing in old magazine articles and even the street addresses of the first 342 people listed in the book *American Men of Science*. They all followed the same law, around 30 per cent of the numbers began with 1, 18 per cent with 2, and so on. For his diligence in documenting where this distribution pops up the law was named after Frank Benford.

This law is yet another facet of scale invariance – one of the signature properties of Universality. It applies no matter what units you use or the scale you choose to make your measurements. As such it gives a clue to where Universality is applicable. The distribution of results it explains only begins to show itself in situations where there are lots of data and there are no artificial limits to the values the numbers can take. It isn't found, for example, in the price movements of America's top ten beers. The sample is too small and prices are kept within artificial limits. The law won't be found in entirely random collections of numbers either. The distribution of first digits in such samples is by definition equal. Instead it applies 'to numbers occupying the middle ground between the rigidly constrained and the utterly unfettered'.

Simple statistical tests will quickly show whether a sample of numbers, say expense claims or sales figures, conform to Benford's law. If they don't it will be very obvious. Because of this it is proving very useful for governments and police forces trying to spot fraudsters. By the nature of what they do fraudsters are tinkering with the normal distribution of a set of numbers. No matter how hard they try to cover their tracks they will not be as good at distributing the results as real life and the normal run of events. It is very difficult to self-consciously mimic these distributions. Unless, of course, they know about Benford's Law too, but we'll leave that distasteful idea aside for the moment.

Thanks to the work by Mark Nigrini, a professor of accountancy at the Southern Methodist University in Dallas, Benford's law is now being widely employed by auditors and tax-collecting services the world over. Already in America it has caught several fraudsters and helped recover or stop the theft of millions of dollars. Now other countries are starting to use it too. Work is also being done to see if it can help spot fraud in other areas, such as the results of clinical trials or in laboratory notebooks, to ensure that any research being carried out is above suspicion.

Firm but Fair

It does not stop there. Universality might also help overturn some cherished notions about the ways that companies grow. Largely thanks to the work of French economist R. Gibrat, economists modelling the growth of companies usually make the following assumptions:

1. The rate of growth has nothing to do with the size of a firm.
2. Growth rates in different years do not affect each other.
3. The actions of one company have no effect on any of the others operating in its sector.

Sadly the theory fares badly when compared with real life. One of its implications is that growth rates are independent of the size of a company. Yet numerous studies of company growth statistics have shown that rates diminish swiftly as the firm grows in size.[19] Large companies never experience again the dizzy rush characteristic of their first successful years. The assumptions also predict that as time goes on there should be more larger companies. Again analysis of the figures shows that the distribution of companies of different sizes remains stable. As it turns out analysis of twenty years of data reveals a power law that does a far better job of characterizing the data.

Luis Amaral and colleagues have looked at data for some of the largest manufacturing companies in the US using data from the years between 1974 and 1993. They used data from the reports that every US company has to submit to the Securities and Exchange Commission that reveals information about sales, assets, number of employees, and such like. To make it easier to compare companies Amaral and his colleagues used only two variables – sales and number of employees – which have the advantage of being measured using the same units (money and people) in every firm. What emerged from the data is that the factor determining how fast a company grows is its size and little else.

This shatters another widely held dogma of economics, that it is the technology used in production that governs the rate at which a firm grows. Better technology reduces the cost of production, leading to efficiencies throughout a company's sales cycle. Obviously the rates apply only within sectors of industry. Growth rates among car makers can't be applied to the pineapple canning outfits of the world. Within sectors though better technology is thought to be the only important consideration. Universality, however, implies something rather different.

Amaral and his colleagues found that large established companies grew at around the same rate every year: a slow steady expansion. Small companies were very different. Newer companies grew faster, but less predictably, than the larger ones. Some burnt bright for a few years, quickly doubling and tripling in size, and then suddenly they went bust.

In any case the key measure of growth was the size of the company, not the technology that it uses. Some of the drag can be offset if a company operates as an independent division within a larger company, but generally the larger the company, the slower it grows. The canonical features of Universality are revealed again as we see that the properties of the elements within a system are irrelevant. In this case size matters. The findings of Amaral and co-workers have been confirmed in a study of data from 80,000 UK companies collected over a similar time period.[20] Gibrat is utterly wrong.

Rather than just stop with this analysis Amaral and his colleagues went further to try and find out if there were other factors determining how quickly a company can grow. Curiously they did. It was not technology but management that dictates the highest rate a company can grow.

Amaral and his team created a computer model of a small company that had a few levels of management between the CEO and the workers on the shop floor. The rules of the model specified that decisions made at one level of the hierarchy directly influence the employees that answer to that manager. The managers and workers were not modelled as mindless peons, however. Instead they were given a 'fuck-up factor' (FUF) which captures the possibility that they would mess up, ignore or otherwise scramble what they had been told to do. A FUF of 0 is anarchy, with everyone ignoring everyone else and doing what they want to. A FUF of 1 produces a robot that slavishly does everything it is told.

When the simulation was set in motion using extreme FUF values the model companies behaved like nothing on earth. However, when the FUF factor was set between 0.7 and 0.9 the behaviour of the model companies was almost indistinguishable from the real ones.

In small companies the FUF can check or speed growth, simply because in a firm that only employs a few people one person has relatively more power. Large companies tend to soak up the fuck-ups. The biggest lesson here, however, is that a certain level of fuck-ups are natural: they are built in from the ground up. The simulation implies that between 25 and 20 per cent of all instructions are messed up by managers and workers. Up to a quarter of all commands will not be carried out as the person issuing them would like. And there is nothing that can be done about it. It's the law.

If Universality can use our fuck-ups to model one company's growth, then it is worth asking just how far beyond the business world Universality can reach and how it touches and shapes the rest of our lives.

Chapter Six

Me and U

How every person is a new door, opening up into
other worlds. Six degrees of separation between me and
everyone else on this planet. But to find the right six
people . . .

Six Degrees of Separation, John Guare

Every psychological pattern is determined; and, within
the cage of flesh and memory, the total swarm of such
patterns is no more free than any of its members. To talk
of freedom in connexion with acts which in reality are
determined is madness.

After Many a Summer Dies the Swan, Aldous Huxley

Disease Dynamics

If you were asked to conquer a nation of 20 million
people how many troops would you need to complete
the job? 100? 100,000? Not sure? Perhaps a few more
details about what you are being given to help you
will focus your thoughts. You have a slight military

advantage over your enemy in that you possess mounted soldiers, steel swords and armour, plus crude guns called arquebuses.

The shock value of the horsemen will be considerable because the people you will be fighting have never seen a horse, let alone faced a cavalry charge. But there your advantages will end. The superiority these weapons confer will be more than compensated for by the sheer number of troops your opponent can bring to the field of battle. To make matters worse the country you are attacking is months of travel distant from your homeland, and, once there, you will not be able to call for more supplies or reinforcements to shore up your position. You're on your own.

The nation you are being called on to subdue is no ragged collection of hunter-gatherers either. It is well organized and its cities are as large as those found in Europe. Its troops are loyal to their Emperor and the whole culture is fiercely militaristic. Years of internal conflict mean that they know how to wage war.

Your answer is probably going to be along the lines of 'As many as possible' or 'Everyone who can hold a weapon'. Yet in the sixteenth century Hernando Cortés overcame the 20-million-strong Aztec nation with only 600 troops. His countryman, Francisco Pizarro, did even better. He conquered the Inca nation, which also numbered in the millions, with only 168 men. At the battles of Jauja, Vilcashuaman, Vilcaconga and Cuzco, Pizarro's Spanish horsemen numbered 80, 30, 110 and 40 respectively. They took on, and beat, thousands, and occasionally tens of thousands, of Indians.

Of course, both had on their side countless millions of microbes that did some of the work of killing the native Indians almost before the Spanish troops could get to grips with them. Those that survived the disease were left demoralized and often leaderless. The Spaniards' conviction that they had God on their side was borne out by the fact that they went untouched by the disease that killed off so many of the American Indians. The native troops crumbled in the face of such moral superiority.

The reason that the diseases the Spanish brought with them spread so fast can be explained using Universality.

Sick Notes

There is no doubt that as mankind has expanded and prospered, disease has done the same. When humanity was split into isolated bands of hunter-gatherers diseases only struck small groups. There was no opportunity for the microbes or viruses to infect large numbers of people because the population was so small and contact between groups was low. As a result only when these nomadic groups disturbed local reservoirs of disease did they suffer infection. Living in, or on the edge of, forests probably meant they regularly suffered from disease. This had an undeniable effect on the course of human history because it meant that there were some places people could not colonize because the microbes lying dormant in the local flora and fauna, until disturbed by humans that is, were too virulent.

It is for this reason that the large herds of ungulates (vegetarian hoofed mammals) of the Africa savannah persist to this day. Many of the species of antelope that graze on the plains harbour a parasite that causes them no ill-health, but when transferred to humans by the tsetse fly causes trypanosomiasis or sleeping sickness. The high mortality rates this disease causes limited the expansion of African peoples. The disease was largely fatal until the 1990s when cures became more available.

Disease only really began to shape human history when it had the means to spread quickly. Diseases such as the Black Death that devastated European nations during the Middle Ages, and carried off about 25 per cent of the population, had been first seen long before. Bubonic plague was first seen in Europe as the Plague of Justinian in AD 542–43. Then it was an isolated outbreak that did not spread beyond a single region. The Great Plague of Athens that was documented by Thucydides in his *History of the Peloponnesian War* struck the city state between 430 and 425 BC and killed hundreds of thousands. It is now thought that it was one of the first recorded instances of an outbreak of the Ebola virus.[1]

The success of the Roman Empire and all the roads it built meant more than just troops could travel quickly; they gave disease an avenue of escape too. The increased movement of peoples, troops and the occasional forced relocation of subdued peoples meant that Roman civilization was prey to occasional epidemics too. One of the first recorded smallpox epidemics occurred between AD 165 and 180. It was known as the plague of Antoninus and killed millions of Roman citizens.

Plague Path

The way that disease spreads is exactly analogous to the way that clouds of magnetism can spread along the length of a magnet held at a critical point. It also has similarities with the way that liquids mix by percolating through each other. A ragged twisting trail of magnetism can easily leap into being across the entire length of a magnet. If you froze the action at that precise point you would be able to trace an unbroken path all the way across it. The ability for changes to spread over arbitrarily long distances is the essence of Universality. At the very least you could follow the route of its influence along the magnet, which also might span the length of the system.

The same can happen with diseases. Once populations become sufficiently large then a phase transition occurs.[2] At this point diseases will be able to span the stretch of the population by a contact path. Above the threshold the disease finds it easy to leap from one city to another or one continent to another. The path might span seas via merchant ships or conquering navies, it might travel along with camel trains or wagon trains, and it might take a long time to travel the route, but the path will be there.

As we saw earlier the dynamics of the critical point and percolation can be shown to be mathematically identical. Recent simulations of the spread of disease have revealed exactly the same fat-tailed distributions of events that are seen in more classical critical systems.[3]

The implication is that we are seeing one and the same phenomenon. The 'fat-tail' distinguishes critical systems from those exhibiting normal distributions where values fall equally either side of the average. By contrast, critical systems have a 'fat-tail' which means they have a disproportionate number of large events.

In the case of epidemics and infections we are not seeing microscopic magnetic grains changing their orientation and tracing out the path; instead we are seeing disease carve a path through a population by infecting people. Some of those infected, but not yet feeling the effects of the illness, will flee the area of infection or be carriers who do not fall prey to the infection. Those running for their lives might, unwittingly, be helping to keep the disease moving and bringing doom to countless others. This has parallels with systems held at the critical point in which small changes can have long-reaching effects. In this case it can be one person acting as a carrier for an infection.

Possibly one of the most famous examples of one person spreading disease is that of Mary Mallon, or 'Typhoid Mary' as she became known. A 1906 outbreak of typhoid in Oyster Bay, New York, was traced back to Mallon who was working as a cook in a Long Island household. She moved on and took up other positions as a cook in several homes in New York over the next few years. The authorities, in the shape of sanitation engineer George Soper, caught up with her in 1907 and she was moved to an isolation ward where she spent the next three years. Soper reconstructed her employment record and found that she was responsible for typhoid outbreaks in almost every home in which she worked. She was responsible for causing the infection of around 1300 people, around 10 per cent of whom died of the disease.

Outbreaks and Epidemics

While these dynamics are similar to those seen in systems exhibiting signs of Universality there are caveats. Diseases that are too virulent and swiftly kill all their hosts will not spread. If everyone is dead there will be no one around to take the microbes on to the next settlement. This has been advanced as one of the reasons that Ebola has not done more damage. The disease has such a high mortality rate that few people are left alive to move on and take it with them. There are also questions about how exactly it spreads from host to host.

Diseases that best exploit the dynamics of growing populations are those that make people sick but don't kill. Diseases that force an immune reaction prosper and persist. Those that do not die, unwittingly carry it to populations that have yet to suffer its depredations. For this reason it is only when populations grew to sufficient size that humanity really began to suffer from outbreaks of disease. Generally it takes a population of around 500,000 people to sustain a disease in this way and ensure there are enough people left alive to carry it on. Any fewer and there will not be enough human fuel to keep the pestilence live.

As a result, during the Middle Ages when trade routes

began to open up and populations began the long slow march from the countryside to the cities, epidemics grew more prevalent. Societies that use agriculture can support much larger populations than those that rely on hunting and gathering. Only when farming took hold and large social groups or settlements emerged did infections become epidemics and start leaping from group to group, town to town.

Bubonic plague burnt its brightest from 1346 onwards, when the opening up of new overland routes allowing trade with China inadvertently opened up a path into European cities for flea-infested furs from plague-ridden areas of Central Asia. The result was the Black Death that killed one-quarter of the European population between 1346 and 1352. In some cities it only left 30 per cent of the inhabitants alive.

The greatest, or worst, demonstration of the devastation disease can cause was during the Spanish conquest of South America. Cortés reached Mexico in 1519 and his first attempt to subdue the nation with a few hundred troops did not end very well. When he tried again with only a handful more soldiers smallpox was already raging in the continent and doing a lot of the work for him. It is thought that the smallpox reached Mexico in 1520 carried by one infected slave from Cuba. The effect was devastating. The epidemic killed almost half of the Aztecs including the emperor Cuitláhuac. The rest were left leaderless and demoralized believing that the Spanish were on a divine mission because they were left unharmed by the plague that was striking down so many of the native Indians. By 1618 the Mexican population of 20 million was down to 1.6 million. It was a shadow of itself and the country belonged to Spain.

The devastation caused by disease did not end with smallpox. If that did not get them then measles, influenza, typhus, diphtheria, malaria, mumps, pertussis, plague, yellow fever or tuberculosis would. Now it is estimated that the arrival of Europeans in America killed off 95 per cent of the indigenous peoples largely because of the diseases they brought with them.[4]

It is sobering to note that international jet travel has interconnected populations more closely than ever before. As a result it has accelerated the speed with which people can move around and take their microbes with them. This is likely to mean that diseases travel faster and infect more people than ever before.

In 1999 and 2000 New York was sprayed with mosquito-killing insecticides because of an outbreak of West Nile Virus fever that has killed seven people. Discovery of mosquitoes carrying the disease caused Central Park to be closed while it was doused with insecticide. The disease had never been detected in the city before and was thought to have been brought into the region by an airline passenger infected in Africa. Now it is spreading to Connecticut and New Jersey. A further fifteen states are on the look-out for signs of the virus which can lead to fatal swelling of the brain in humans.

As you would expect the last one hundred years has seen some very large epidemics. The 1918 flu outbreak can be traced to the fact that European, American and African troops enjoyed each other's company in northern

France at the end of the First World War and when they parted took back with them new strains of microbes they were previously immune to. The outbreak of influenza that followed swiftly overwhelmed medical facilities largely because many of the countries it struck were still recovering from being war economies. It is estimated that 20 million people were killed by the flu outbreak.

In 1957 a strain of 'Asian' flu threatened to wreak the same havoc but rapid development of a vaccine checked its spread. Now only immunization programmes and modern medicine are stopping diseases, both old and new, from spreading swiftly and infecting huge numbers.

Social Statistics

Diseases can spread in this way because of the physical connections that link societies and individuals. In the Roman Empire it was the network of roads fanning out from the heart of the empire, but in modern times it is international air routes that help us travel, mingle and spread disease. However, the connections between people are not solely physical, they are social too.

There is no doubt that we are tied into huge social webs very tightly. These connections, again thanks to an increasingly mobile population, are becoming ever more inclusive. Most people work in a town other than the one in which they were born. Many people have gone to university or college outside their hometown. Their parents usually live in another part of the country, and the roots of their grandparents are probably fixed even further afield. As we expand and our personal connections mesh the social world shrinks.

In the 1960s Harvard University psychology professor Stanley Milgram was one of the first to study the reach and intricacies of such social networks. Milgram showed his characteristic ingenuity in the method he picked to study social connections. He took copies of a letter he had written to a stockbroker friend of his in Boston and distributed them to a random group of people in Nebraska. The letters had to get to the stockbroker but only by being passed from person to person. They could not be posted. Milgram stipulated that the letters could only be passed on to someone that the passer knew on a first-name basis. You would think that this would mean the stockbroker never got to read what Milgram had written to him, and yet most of the letters sent this way reached the Boston stockbroker. They did so only after passing through the hands of six people. It is from this that we get the notion that we are only 'six degrees of separation' away from anyone else on the planet.

This phenomenon has now become a play, a film and a parlour game. It has even assumed new variants such as Six Degrees of Kevin Bacon, as well as others involving baseball stars and even Monica Lewinsky. The aim of the game is to get from Kevin Bacon or Monica Lewinsky to another person in as few steps as possible. Kevin Bacon was picked because he has appeared in so many films that it is usually easy to link him to almost any other actor or actress in only a few steps. Getting from Kevin Bacon to Kevin Costner takes just one step because both were in Oliver Stone's *JFK*. Most actors, living or dead, can be connected to Kevin Bacon in only two or three steps. It takes very obscure actors, such as Sri Lankan actress Grace Ariyawimal who has only been in one film, to be more than four steps away. The game was invented by Pennsylvania college students Craig Fass, Brian Turtle and Mike Ginelli. For obvious reasons Six Degrees of Monica Lewinsky is not played with actors or actresses. Instead the aim of the game is to link Ms Lewinsky with American political figures usually by as salacious a connection as possible.

Memes Away

If social connections can be mapped in this way it might be the case that ideas travel like diseases and spread like viruses. Memes may spread in the same way as an infection. A meme is like a gene but one that is preserved mentally rather than in DNA. It is a gene for culture, and culture has many of the attributes of a system displaying Universality. It is open and information flows easily and constantly across it; it is made up of billions of individual elements, you and me; and tuning of a sort goes on in the form of marketing, advertising and political propaganda.

Certainly the environment that shapes us both physically and mentally takes in a very broad sense of the word. Just as childhood diseases such as chicken pox can leave marks that persist throughout our lives, so perhaps memes that we catch in early life can stay with us too.

Environment includes the physical world and the cultural one, the realms of both objects and ideas. As well as the fields you played in and the food you ate, it includes the books or comics you read as a child, the social milieu you were brought up in, the TV programmes that you watched and the music you listened to. These articles of consciousness can have as profound an effect on shaping a view of the world as growing up in Yorkshire or Yemen. As Simon Schama says in his magisterial work *Landscape and Memory*: 'For although we are accustomed to separate nature and human perception into two realms, they are, in fact, indivisible. Before it can ever be a repose for the senses, landscape is the work of the mind. Its scenery is built up as much from strata of memory as from layers of rock.'[5]

Environment then is as much the mental as it is the physical. The choices available to children in 1899, 1999 and 2099 will be very different. The prevailing opinions of a society are like a national park. They have ideas that are as hard to scale as rearing cliffs as well as those that are like paths that are easy to walk along.

Perhaps memes need a threshold of people to carry and propagate them just like viral genes. It might be the case that the plethora of ways that our cultural world is connected – talking, radio, TV, cinema, music, books and the Internet – means that it is held at the edge of chaos. This might explain the occasional success from nowhere experienced by bands such as Hootie and the Blowfish or by authors such as Dava Sobel. As we become ever more connected by technology, be it jets or e-mail and telephones, it gets easier for information to spread an arbitrarily long way. This again is one of the signature properties of Universality.

The influence of memes can be felt over long periods of time and still perform a function even if they lose some of their original meaning. Just like DNA. For example, the golden age of the British navy fell during the late eighteenth and early nineteenth centuries. Today the Royal Navy is a shadow of what it was then, and Great Britain does not depend upon it for its standing in the world. Yet the English language is peppered to this day with phrases that date from this era and many people are unaware of their meaning. Phrases like 'first rate' or 'sixth rate' were used to describe a ship based on the number of guns it carried. More senior captains got a higher-rated ship. Now the words are used to mean either a good thing or a bad thing respectively, but few people know their original meaning.

The phrase 'to the bitter end' is another example. Most people assume this means until a distasteful conclusion but the name comes from the bitter, or inboard, end of a cable attached to the anchor. On a ship the main stay was one of the strongest ropes holding up the main mast. It retains some of its original meaning to this day. Similarly the phrase 'the devil to pay' sounds straightforward enough but again has a maritime origin. The 'devil' was the seam between the deck and the hull of a ship. 'Paying' meant sealing it with pitch but the narrow angle between deck and outer boards meant it was hard to get at and a difficult job to do.

There are other examples too. People who are a little crazy or foolish are sometimes described as 'doolally'.

The phrase actually comes from a staging camp in India called Deolali, where soldiers or civilians on the way home would wait for a ship to carry them back to England. Often they had a long wait and under the fierce heat many suffered sunstroke. Anyone falling victim to this was said to be suffering Deolali Tap. Now the phrase has been corrupted to become Doolally tap, or just plain doolally.

Like junk DNA these memes persist even if few people recognize them for what they are and they no longer perform their original function. But they still have their uses. It might be that the same dynamics seen in evolution, which have a lot in common with Universality, are at work in ourselves and the culture we create. Many of the theorists working on complexity theory have speculated that events in history follow a pattern that is best understood as a dynamic physical system, one that resembles a system at a critical point. It is hard to see how that could not be the case. So many systems do exhibit these dynamics that we would have to claim a very special place for ourselves to avoid this conclusion.

There is no lack of historiographers who have tried to fit patterns of one sort or another to the sweep of events. Arnold Toynbee saw civilizations passing through four phases: an age of growth or expansion; a time of strife and unrest; a universal state; and then a final pause or interregnum before collapse. He saw this repeated over greater and shorter time scales among many different civilizations and spent much of his life finding evidence to back it up.

Nazi historian Oswald Spengler believed that civilizations develop like plants and have their seasons of growth. During their spring civilizations flourish and expand rapidly. In their summer they establish themselves, and a culture of manners develops. Their autumn sees cities and monarchies spread wide, and with winter comes scepticism, materialism and idealism. After this cultures die off to be replaced by new growth. Again Spengler saw this happening to both great and small nations and across time. By contrast Hegel saw history as something powered by the conflict of ideas with the emergence of successive philosophical movements as driving progress. For him history was overlapping waves of ideas, each one of which pushed a culture along.

Many people have tried to fit all kinds of explanations to the playing out of events that is history. Perhaps the eternal debate over whether we are doomed to repeat our history, sometimes as tragedy and sometimes as farce, shows that the evidence can be made to fit almost any pattern you care to squeeze it into. If it is the case, however, that history is another critical phenomena, then you should be able to pick almost any scale and find significant events. There are events of all magnitudes at all scales after all. The bigger changes may be more evident but they are not necessarily more important, nor do they have as far reaching an impact. Certainly there is no doubt that small events can have very long-range consequences: think of the assassination of Archduke Franz Ferdinand and his wife and all the lives lost as a result of that single act.

Crowd Trouble

It is not just larger social movements that resemble a critical system. If you get enough people together then the pattern of behaviour that results has a self-similar cast to it. We flock in fractals. Gather a crowd, get it moving and a fractal order will emerge. Whether the people are calmly exiting a rock concert or struggling to escape a burning building there are strong regularities to be seen in what they do.

Keith Still is the man who has spent over eight years obsessing over the fractals submerged in crowds. His interest dates from 4 p.m. on 20 April 1992. Still had been waiting to get inside Wembley Stadium in London for a rock concert for over four hours. To pass the time he watched the parts of the crowd that were moving and saw some complex patterns of movement emerging. Before Still started doing his work crowds were modelled using the same techniques that were applied to turbulent liquids. What Still saw convinced him that this approach was mistaken. The crowds weren't acting like liquids at all – there was something else going on.

He realized that the dynamics of a crowd were determined by the trajectories of each member. We trace out a pattern based on what we are trying to achieve. In a nightclub it might be trying to get close to the bar or someone you want to dance with; in a rock concert you might be trying to get closer to the band or away from the dangerous dancing going on in the mosh pit in front of the stage. The speed at which we achieve these goals is determined by the speed we can move through the crowd and the physical layout of the place. Still found that although interactions between individuals tend to be local, i.e. we take into account people immediately around us, the effects of movements permeate the crowd. As people move they create a 'wake' that tends to pull in other people and drive the structure of the crowd. Once you mix enough people trying to move you get fractal patterns as these strategies and goals collide and interact.

Still used his original work to model how people escape burning buildings. The early simulations were built up from watching what people do in real emergencies. Curiously many people do not panic in such situations. Most are happy to follow the crowd as it makes its way out of a building. Few act like anything unusual is happening.

Men are more likely to investigate fires before deciding what to do, whereas women tend to want to get away as quickly as possible and follow the crowd. When people are making their exit they tend to head for the one they most usually use rather than the one that is closest and will get them out of the building quickest.

Still has now formed a company called Orchid Fractal Engineering that is turning his work into simulation tools to help architects design buildings that are easy to escape from in the event of an emergency. Already the modelling system, called Legion, has been used at Wembley Stadium to improve the way that crowds can leave the area and it is helping to work out the best way to site ticket barriers at busy train and underground stations.

Jam Today and Tomorrow

This fractal organization is evident even when we get off our feet and into our cars. Traffic jams on major motorways are another fractal phenomenon. As everyone knows driving a car along a motorway involves constant adjustments for other users of the road. Those going too slow have to be passed, those hogging the middle lane have to be overtaken, and those speeding like loons have to be waved on by.

This interaction creates a complex structure and on days of heavy traffic shows all the characteristics of a system at a critical point. An accident on one side of the carriageway will create a backlog of traffic on that side. On the other side rubber-necking by drivers creates a phantom traffic jam. This small change ripples back down the traffic as drivers slow down so as not to crash into the cars slowing down to have a look at what has happened or slow down because the person in front is putting on his brakes. A small change has immense consequences far from its starting point and its effect is felt for a long time to come. No one is in charge, but everyone is late.

The UK Department of the Environment, Transport and the Regions is experimenting with phased speed limits in a bid to make these phantom jams evaporate.

The department reasons that if it can slow down the traffic before it reaches the phantom jam then there will be less risk of vehicles getting caught up in it. Of course, the danger is that they may actually create more jams than they are stopping. The department claims that by using these methods they can increase the throughput of traffic fourfold. Whether their efforts do what they want remains to be seen, but next time you are on the motorway and stuck in a traffic jam that appears to have formed for no reason you will at least have a better idea about what was going on. However, it is unlikely that this knowledge will be of much comfort.

It is not just on physical roads that these patterns can be seen. The virtual lanes of the information superhighway show the same regularity. Bernardo Huberman and his colleagues from the legendary Xerox Palo Alto Research Center, where they developed the first windows-based computer in 1972, have found similar distributions in the way that people surf the World Wide Web.

You would have had to have been dead or locked in a box for the last few years not to have noticed that the Internet is becoming vital to many people's way of life. The numbers estimated to be online at the end of 2000 are just under 410 million people, and the flocking of people to the Internet looks like it will continue for a while yet. When all these people go online they seem to search in the same way. By looking over the electronic shoulders of 23,692 users of the America Online network Huberman has found that we tend to approach the finding of information on the Web in the same way.[6]

The researchers looked at how deeply surfers go into a website before getting bored, finding what they want or giving up and going elsewhere. Typically people are prepared to look at three pages, or fewer, on one website to find what they want. They give up quickly if it will take longer.

This research reveals that there is a Zipf-like pattern to the way that people search and surf the Internet. George Zipf discovered that the distribution of words in documents as disparate as daily newspapers and James Joyce's *Ulysses* followed a set pattern. The same is true of information on websites and the patterns of surfing the sites themselves. Most people stick to the sites that are easy to find and rarely venture beyond that. Just as in writing most people make far more use of words such as 'the' or 'and' than they do 'snot-green', and very few would use 'agenbite' or 'inwit'. The phrase 'agenbite of inwit' is one of the most famous phrases from Joyce's *Ulysses*. It is used to describe the internal pull of conscience and regret felt by Stephen Daedalus, one of the main characters in the book.

Clap Happy

Applause also seems to be a critical phenomenon. Research by Albert-László Barabási, associate professor of physics at the University of Notre Dame in Notre Dame, Indiana, has shown that once applause gets going its dynamics become very sensitive. While the volume and heartiness of applause may be mediated by cultural differences the way it starts, spreads and sustains itself is not.

Barabási and his co-workers recorded, digitized and analysed the applause given after performances of opera and stage productions in theatres in Hungary and Romania. Recordings were made both by a microphone hanging from the ceiling of the concert hall and others placed near randomly selected individuals. They also conducted controlled clapping experiments with seventy-three high-school students. One sore-handed individual had their clapping patterns analysed in a series of experiments conducted over an entire week.

The researchers found that the spatterings of early applause quickly coalesce into a wave that takes everyone with it. After a few claps at different speeds people tend to catch on to the clap rate of the whole audience and applause becomes sustained and cooperative. This is a little like the sensitivity of a system at a critical point to small changes.

Curiously if any opera or theatre-goers are unhappy with the volume of the clapping and don't think that it expresses how they feel there is little they can do about it. They cannot increase the amount of noise that they are making so they tend to simply increase the speed at which they clap. If enough people do this the coordination falls apart and the applause gets ragged and chaotic, just like a critical system that is pushed beyond the point at which Universality can be seen. The sensitivity to change and the self-similarity disappears and it descends into white noise.[7]

Even the performance of athletes looks like it obeys a power law. Races at international athletics meetings have been timed for over a hundred years and the information about who was fastest over which distances represents one of the longest, most reliable and up-to-date indices of human endeavour in existence. By analysing the data from years of athletics meetings and swimming contests Sandra Savaglio and Vincenzo Carbonet have found that human performance in these events follows a power law.[8]

The pair decided not to look at the shortest spring races because in races around 100 athletes barely breathe and burn energy anaerobically. By contrast in longer events aerobic respiration is all that matters. As anyone who has run a marathon or half-marathon will testify you breath a lot while you are doing it. Events that take longer than about 150–170 seconds to complete will involve a switch from one type of respiration to another. Their research focused on longer events to be sure like was compared with like.

The results show, perhaps unsurprisingly, that it gets harder to break records as more and more races are held. Large earthquakes are rare and so are athletes of the calibre of middle-distance runner Steve Ovett or Ethiopian long-distance champion Haile Gebreselassie.

Folk Laws

In groups then, socially and physically, there seem to be some things we all do in the same way that build up to create these patterns. They are more evident in large bodies of people but, in some circumstances, we still do them when we are alone. There is no doubt that there are some things we all do in the same way. We tend to stick to familiar territory and, in our routine activities such as shopping and socializing, we tend not to stray outside those areas. We regularly go the way we know and it makes sense to do so.

In stressful situations, however, when we have to react quickly and don't have routine to guide us, differences and preferences assert themselves. For instance, women tend to head uphill when lost whereas men go downhill. Right-handed criminals generally flee to the left, but avoid obstacles by going right. They also stick to outside walls when trapped in large buildings.[9]

People do react differently in these situations, but they tend to pick from a limited repertoire of responses. In both situations there are strong regularities to the decisions we take and how and where we spend our time. If we watched ourselves for any length of time we would find that we create an 'activity space', a bubble that encompasses all the places we go and what we do.

Such behavioural quirks are helping the police catch criminals. Canadian criminologist Kim Rossmo has developed a computer system called Orion that knows the geographic tendencies of people and uses this to plot where criminals might live and what they might do with their victims. Consumers and pathological murderers seem to apply the same rules when selecting a hairdresser or a place to bury a body. Orion uses rules garnered from years of police work, and uses them to plot the likely places that a criminal might have come from.

The computer system is proving enormously useful. Rossmo is keen to stress that it cannot catch criminals but it can help detectives cut down the places that they have to look and who they have to question. Already it has been instrumental in helping police catch sex offenders, murderers and arsonists.

The question now has to be asked if we are scored through with self-similarity too? There is no doubt that fractals are key to a huge amount of human endeavour. Zipf showed that language is pitted with them, and power laws have been found at work in some of the greatest works of music. They have even been found in the abstract expressionist paintings of Jackson Pollock.

Many of the huge paintings that Pollock produced between 1943 and 1952 were created using cans of dripping paint. Pollock often worked on them for months gradually building up the complexity of the image. Typically the canvas lay on the floor of his barn and Pollock suspended paint pots over it and let the paint drip as the pot was pushed to and fro.

You might expect that this method would produce nothing but a mess, but many of the canvases have a subtle symmetry that stops them being mere daubs. Pollock himself thought that many of these works represented 'pure harmony' and now science has proved him right.[10] Even though chaos theory and fractals were unknown outside science when Pollock was creating these paintings, he knew what looked good and what would rise to the eye.

All these things might be the external manifestations of the fractal shape twisting, beating and pushing at our core. It might be of a slightly different dimension for everyone but it looks like we all share it. We know it when we see it.

Voodoo Child

It might be the case that we have less choice than we think in many of these situations because it is becoming obvious that a lot of what we do is not under our conscious control. Many psychologists have started talking about the zombie that is lurking within us all. This non-conscious self takes care of the day to day business of being a body. It looks after all those things that need no conscious mediation. The things you do without thinking. When you get to the bottom of the page and turn over the zombie will be the one in charge of marshalling your muscles to do it. You don't have to tell your arm what to do. Though you probably will now that I have mentioned it.

The autonomic nervous system looks after a lot of basic bodily functions but the zombie is supposed to be a little more cerebral than that: not as cerebral as doing algebra but more so than keeping you breathing. It may be just a metaphor but it captures the essence of a bizarre division in our mental lives.

The zombie looks after phobias and has your heart

pounding even before you have really seen the spider in the bath or the daddy-long-legs clattering around the room. It is far harder to fool the zombie with optical illusions. Take two wooden discs of identical size. Place each one in the centre of a circle of discs. One circle should be made up of discs bigger than the original two discs. The other circle should be made up of smaller discs. Although you know that the two central discs are the same size, one will look bigger than the other. The zombie won't be fooled, however. If you are asked to reach for the blocks your hands will open just as wide for each one.[11] The zombie is also the one that niggles you with a hunch. In those unexplainable situations where you find yourself convinced about something for reasons that you cannot explain, it is the zombie that worked it all out. It is obviously capable of making value judgements and weighing evidence.[12]

The concept of intuition pops up in almost every culture and goes by many different names. In English it is called *nous*, in French *coup d'oeil* and in German *fingerspitzengefuhl*. Philosophers such as Henri Bergson, Baruch Spinoza and Bertrand Russell, as well as many others, all distinguished between knowledge we learn and that we acquire.

Gary Klein, an Ohio-based psychologist, has spent the last fourteen years studying the informal expertise that professionals such as firemen, nurses, air traffic controllers and music teachers pick up on while doing their job. He has documented thousands of cases where a gut feeling about something or someone has proved to be right. He has found that this intuitive intelligence works best when people are under stress: when information is incomplete and they have to make a decision quickly. Klein believes that the professionals are nonconsciously picking up on clues that only their expertise alerts them to, and then using this to make their decisions.

Further research by psychologist Joseph LeDoux has found a pathway in the brain that stores some memories in places that cannot be accessed by conscious thought. He believes that we call on these when we act on intuition. Decision-making is a compromise between our conscious thought processes and nonconscious intuition.

History Man

The evidence suggests that we create our view of the world as much as we react to what is actually out there. Certainly it is true that we construct a lot of our world rather than simply react to it. Anticipation is useful in pure motor skills. We often have to catch things that are moving faster than we can follow them and the ability to work out where they are going to be allows us to put our hands in the way and pluck them out of the air. This anticipation also overflows into other areas of our lives. Take a walk through a dark wood on a dark and stormy night and see what your brain can conjure up out of the shadows. The brain does not rely on just the wider world or inner space. Its, and therefore our, reactions are produced by an interaction between the two.

A comprehensive study by psychology professor Dennis R. Proffitt from the University of Virginia found that a person's view of the world is coloured by how fit, encumbered or old they are. Fit people tend to view hills as flatter than they actually are because it is easy for them to scale the inclines. Elderly or unfit people, and even those carrying heavy weights, tend to overestimate the effort it will take to reach the summit of the hill. This implies that the way that we can move around the world colours our perceptions and leads us to interpret the world in a way that reinforces our impressions. Unconsciously we construct the world to fit ourselves.

People often have to rationalize the decisions that they take. Studies of people who consider themselves lucky or unlucky show that what we call good fortune is often an after-the-fact explanation of a decision we always take. It helps if you are an optimist too. Luck has more to do with how you explain what you do to yourself than anything else.[13]

Some classic experiments conducted in the late 1950s show evidence for the same phenomenon. In the experiments psychology professor Donald Hebb and colleagues from McGill University in Montreal tried to find out what happened when people were deprived of normally varying sensory input. If we are nothing but the reflection of the world around us then one would think that subjects deprived of input would spend their time asleep or very bored.

Hebb and his colleagues had to pay subjects to take part once they knew what was going to happen to them. The subjects were made to lie on a comfortable bed in a quiet cubicle. They had to wear goggles that diffused the light entering their eyes preventing them from picking out patterns or recognizing anything. The subjects were made to wear cardboard cuffs that extended beyond their fingertips. These allowed them to move their limbs but stopped them feeling anything with their fingers. As much as was possible the outside world was cut off. The psychologists conducting the research would periodically ask the subjects questions to see if their perceptions or mental abilities were being affected.

Subjects lasted from two to six days in this state of enforced and isolated relaxation. None of the subjects enjoyed their time in the cubicle. Once they had caught up with their sleep and spent time thinking about their lives they got very, very bored. The answers they gave to the questions asked by the experimenters showed that as time wore on they found it harder and harder to concentrate and solve the simple puzzles that they were set.

Many of them started hallucinating. At first they saw geometrical patterns such as flashing lights or moving bars; some saw wallpaper-like designs. As the days wore on, for some the hallucinations grew in pitch and intensity. For a very few people there were entire scenes, small dramas cooked up by their own brains, that had the feel of a dream. Others suffered auditory and bodily hallucinations. The vividness of the hallucinations bothered some of the subjects who complained that they were more real than the perceptually deprived existence they endured during their time in the cubicle.[14]

Further tests administered after the subjects were released from the cubicles showed that their perceptual and intellectual abilities had been briefly lowered.

Who are U?

The picture that is emerging is that the zombie makes a lot of decisions for you. It knows your capabilities and helps you see the world in those terms. Perhaps many of the choices that we make as we grow up are actually made by the zombie. As a result we see what we expect to see and find it hard to escape ourselves and our expectations. Who we are and what we can do colours the way that we see the world. It is reasonable to expect that both physical and mental qualities are taken into account when we are building up this view of life. It also implies that we broadly deal with all the events that life throws us in the same way. We approach them in the same fashion, extract the same meanings from them and reinforce our own world-view with them. This might be the way that the fractal pattern of fate plays itself out.

There is a lot of evidence to suggest that the brain is a canonical critical system. It has all the properties of a critical system.[15] It certainly has enough interacting units. There are up to 100 billion neurons in the brain. All of them are heavily interconnected and become more so as the body carrying them around ages. In this sense we really do become set in our ways. There is no doubt that the brain, like the lungs, is arranged fractally. Only by nesting and knotting neurons can so many of them be crammed into your cranium. If you removed a human brain and flattened it out it would be as big as a tablecloth. That's more than enough storage space.

There is also a constant flow of energy and information across the brain. In fact the brain consumes 20 per cent of all the energy you take in. A lot of blood has to flow around it to deliver these nutrients, so much so that the ancient Greeks thought that it was a kind of radiator and its only function was to keep the blood cool.

As well as energy, information is constantly passing across it too. The rhythms of the brain are constantly driven by what the body senses. This sensory information reaches it in astonishing amounts and it soaks up every last morsel. At the same time the body is constantly telling it how it feels and bathing the brain in a variety of hormones that colour its mood and partly determine how it reacts.

Cerebral Pulse

There are more direct physical measures of the brain that imply that it displays Universality. The speed of thought is a leisurely few metres per second. There is a measurable delay between stimulus and response especially if the longest neural pathways are involved, such as the ones connecting your toes to the sensory centres in your brain. If someone sticks a pin in your little toe while you are not looking it will take a moment or two for you to react. If they keep doing it the delay will become easier to notice because nerves take time to recover and get ready to react again. It is as if there were only one messenger for each neuron between toe and brain. The message telling your brain that the toe is being tormented is carried by this messenger, who then has to return to the start of the nerve before he can pick up another message. Hence the delay. This is why people who have suffered traumatic injury say that pain comes in waves. It does, because it takes time for the nerves to recover. Between the messengers rushing along the nerves there is a short moment of respite between each one.

When the stimulation threshold of a neuron is overcome either by direct stimulation, such as a pin, or the summed signals of other nerve cells, a signal streaks down the length of the nerve so it can pass its message to the neural tissue beyond it. The signal that is set off takes the form of an ionic exchange in which sodium ions flood into cells. Crudely this causes a chain reaction that sends the signal shooting off down the rest of the nerve. Before the nerve can react again the excess of sodium ions has to be pumped out. As a result there is a latency period between reactions. This is the speed of sensation and thought.

It has long been known that there are waves of activity in the brain. The anatomy of the brain is divided by function and it is thought that these waves help synchronize the disparate parts of the brain working on a common task. Brainwaves are differentiated by their frequencies. Delta waves, which have a frequency of less than one cycle per second, occur when someone is asleep. They are evidence of the brain just ticking over. Alpha waves occur when people are relaxed and are not concentrating on anything in particular. Faster beta waves, which have a frequency of up to 25 cycles per second, appear when someone is doing mental arithmetic or is concentrating on a problem.

Waves of excitation spread across the heart muscle cells in the same way and in both organs this fire–reset–fire sequence creates a fractal signature. In a healthy brain the magnitudes of these sweeping waves as they racket around the brain follow a power law.

This Island Order

Physicist Peter Jung, working at the Georgia Institute of Technology, has done research on a part of the brain of one-day-old rats called the hippocampus. This structure is part of the limbic system and is involved in learning and the formation of memories. He found that the waves of communication flowing through the hippocampus followed a power law.[16] Although he was working with just one of the many types of brain cells most neurons pass on information in a very similar way.

Types of brain cell are distinguished by the number of connections and interconnections they have with other neurons. Motor neurons are relatively simple and have only a few connections to muscle cells at one end and a few hundred connections at the other end where they receive information from the nervous system. The Purkinje cells of the cerebellum, where you do your deepest thinking, can have up to 10,000 connections.

Further work by Jung has established that people suffering epilepsy produce patterns of excitation that are very different to those seen in healthy brains.[17] It has also been shown that the brainwaves of people in a coma are very different from those of people awake and walking around, and even from those of people who are asleep.

Biophysicist Scott Kelso from the Florida Atlantic University in Boca Raton has found very similar dynamics in human brains rather than those of young rats. He believes that they indicate that the brain is constantly undergoing a series of phase transitions as it organizes reactions and coordinates bodily activity.

Kelso has been measuring mental activity with a superconducting quantum interference device, or SQUID, a sensor that can discern tiny changes in electrical and magnetic fields. The most sensitive versions, cooled with liquid helium, can detect changes in magnetic fields as small as one-billionth that of the earth. The SQUID produces a very detailed picture of developing mental activity. It shows which parts of the brain are involved in coordinating bodily movements and reveals the patterns of neural activity flowing through them. The same kinds of patterns seen in the brain organizing bodily behaviour are also seen in critical systems.[18]

Kelso and his co-workers have found that our coordination and ability to follow and match simple rhythms improves dramatically when the dynamic patterns of activity in the brain undergo a phase transition. This coordinating ability of the brain seems to improve best when islands of stability emerge from the rush of sensory information. These serve to tell the body what it should be doing. Unsurprisingly the self-same dynamics have been found in many other systems. It looks like they are another facet of the principle of Universality, even if they do emerge in a slightly different way.

The dynamics that Kelso and his colleagues have found have affinities with the pioneering work that Mitchell Feigenbaum has done on systems heading towards chaos. Systems such as turbulent fluid flows which look like they are nothing but a rushing chaotic flood, but Feigenbaum found hidden within them order and structure. He found that fractal regions, islands of order, can crop up in a huge variety of systems.

Feigenbaum's insight grew out of the work that Australian physicist turned mathematical biologist Robert May, President of the Royal Society, did on fluctuating populations of organisms. Before May's work it was generally accepted that the populations of organisms in any particular habitat, minnows in a pond for example, would eventually reach equilibrium, Mother Nature and the beasts working in harmony. Ecologists studying populations expected to see a linear relationship between the population one year and the next. A pond can only support a certain number of minnows. Once this limit is reached the population will neither grow nor fall. Barring the sudden appearance of a heron or two that eats lots of the fish the population will stay constant.

Feigenbaum Found It

In linear systems there is a direct relationship between the energy you put into a system and the effect it has. With a small push your sister on the swing will hardly move, a bigger push and she is starting to enjoy the ride, a huge push and she flies off the swing and lands on the lawn. Input (push) and output (swinging sister) are easy to relate to one another. The same push of the same sister on the same swing will produce the same ride. In nonlinear systems, however, the difference between input and output can be huge. Imagine giving your sister the tiniest push with your finger and seeing her fly over the trees at the end of the garden. The output here is many times greater than the input, and this takes a lot of explaining.

When biologists and ecologists, aided by Robert May, began modelling population growth they got a bit of a shock. Equations created to describe population growth have to take into account several real-world facts. The rate of growth they expect is capped because the minnows only have one pond to sustain them. They also have to reflect competition for food among minnows and from other organisms. A small population of minnows will grow rapidly, but if there are lots some will starve. They also included variables for natural death rates and for deaths caused by predation, by heron or pike perhaps.

These equations can be used to predict what next year's population is going to be by using this year's total as the input. Repeat the calculation over and over again and you have a map of the way a population grows. The most important parameter is the one that represents the growth rate. This captures how fecund the fish are or how well they exploit the pond they call home. Lower values of this parameter mean the population will stop growing beyond a certain point; but higher values do not always mean it simply settles down at a higher level. May found that with some values of the parameter the population numbers start to fluctuate wildly, booming one year and busting the next in a seemingly random manner. The populations do a crazy dance between extremes.

Plotting these numbers on a graph revealed a bizarre pattern. It showed that at certain parameter values the equilibrium level of the population fluctuates between extremes on a regular basis, every two or four years. This means that tiny changes in the fecundity of a population of minnows might have a dramatic effect on their numbers.

Increasing the parameter value further reveals greater fluctuations on both longer and shorter time scales. Splits upon splits upon splits. Eventually the population falls into chaos, where no pattern can be seen. Between steady growth and chaos is the region of regular bifurcations, a fractal. Between the regularity of steady population growth and the chaos of an organism struggling to hang on is a region of calm and order. A tea party in the midst of a tsunami.

Building on this work Feigenbaum realized that this type of dynamics was quite common. He found islands of stability popping up in hundreds of places. Many real-world structures and the changes they undergo can be explained this way. Clouds are a good example. Work by Sean Lovejoy has shown that they are fractal over seven orders of magnitude. They form when warm moist air rises and expands in the lower pressures prevailing high above the ground. They are most prevalent along boundaries between low- and high-pressure fronts, between regions of steady change and storms. Rain seems to fall in fractal bursts too. There is talk of using this to try and predict where acid rain might form and help to understand how humans can change weather patterns and perhaps limit some of the damage we do.[19]

In work done at the Los Alamos National Laboratory from 1974 onwards Feigenbaum has found these islands of order in a huge variety of systems: in smoke curling up from a cigarette, in the movement of stock market prices, in turbulent liquids. It seemed to be everywhere. In the jargon of mathematics these patterned points are called *strange attractors*. Drawing a graph of the behaviour of one of these systems often reveals a point about which the system revolves. It assumes a cycle that has a fractal structure that does not reveal itself in any other way.[20]

Feigenbaum's discovery is an adjunct to Universality. It reveals that the kind of order we see in systems at critical points can arise by other means. But the mathematics used to understand them, Wilson's renormalization group theory, is the same. They behave in the same way and sit at the same point between order and disorder. Drive a system that is revealing these islands of order with more energy and information and it will collapse back into unpredictability. Keep pushing it though and the islands will emerge again and again and again.

Moving Target

Now Kelso and his colleagues have found these spots of stability in the brains of people carrying out all kinds of tasks. The brain appears to be peppered with them when it is helping to coordinate motor tasks. The islands of order turned up when Kelso and his colleagues were testing the perceptual abilities and reaction speeds of subjects. The islands stood out in large-scale SQUID measurements of brain activity. The ordered patterns also appear when a person is set a task they have no experience with, such as balancing a pole on one finger. The islands of order appear at all times and frequencies and pop up all over the cortex.

Kelso theorizes that using these emergent patterns the brain manages to organize and coordinate systems that would otherwise be competing. The patterns have the advantage of regularity but they are also flexible and adaptable. As a fractal, the pattern of events – in this case neurons firing in patterns to control muscles – operates on many scales. This makes it easy for the brain to quickly tune its instructions. The implication is that the brain exists on the edge, so it can be flexible and quickly adapt to a changing situation. This region of control is a narrow one, however. Push it too far and its ability to cope with change collapses.

Psychologist George Miller identified the brain's information-processing bottleneck in the mid-1950s. In a famous paper entitled 'The magic number seven plus or minus two',[21] Miller described a series of experiments showing that people can only handle around seven pieces of information at any one time in their short-term memory. The most adept can handle nine and the least adept five – hence the title of the paper.

The seven pieces can be more than single numbers or words. By 'chunking', i.e. treating numbers you are given to remember as blocks or words as phrases many more can be held in your head at any one time. Instead of being able only to remember 1, 3, 4, 5, 6, 7, and 9 you could remember 235, 765, 729, 398, 630, 329 and 917. The 7 (±2) rule still applies though. If you try with more than seven chunks you will start forgetting them pretty quickly.

Miller found that this ability to remember 7 (±2) chunks of information does not degrade gracefully if you have more than that to think about. Any more than 7 (±2) and you have to work harder and harder to hold the information in your head. It is like keeping plates spinning on poles. Too many and you simply cannot get around to them all in time. Soon you will stumble mentally and lose the lot. The result is that briefly you can't hold anything in short-term memory while you recover your wits. It might be that Kelso's islands of stability are at work here helping you keep the chunks of information in mind.

If it is the case that the brain is at a critical point, and it is by no means proven, then there would be a self-similar pattern to the many waves of excitation running through it. Self-similar means what it says: on all scales it looks the same. This might mean that each brain has a signature – a fractal stamp dictating how it will react to different events. It might mean that within the broad boundaries set by our genes and upbringing we always react in the same, fractally filtered, way. In effect the brain will always react the same way in every situation from making a cup of tea to picking a partner. As careful or as careless as you are with one, so may you well be with the other. This is why fate might be fractal. The signature state of our brain might dictate how we react to almost every situation.

This is not fate in the Greek tragedy sense of the word. We should not be looking over our shoulder to spot one of the three Fates who arbitrarily control our birth, life and death. Nor is it fate in the biblical sense of the word. It should not be taken to mean that God is in charge and has set this whole thing in motion, leaving us to play out parts we have no control over. It is not fate in the sense that everything we do is determined from the day we are born.

The fate implied by Universality is much more personal. It is about you and the way you engage with the world. In every critical state a self-similar pattern is evident. So, the way that you go about your life might be subject to the same dynamic and produce a similar pattern. This takes nothing away from the complexity of human life and does not imply that people merely repeat a few simple strategies to get them through the days and months. Fractal patterns are always knotty and complex, and so are your reactions to the world. The boundaries that this fate has to play with are set by our biology and upbringing. They are wide but have their limits. Sadly, we cannot be anyone we want to be. We cannot remake our genes to make ourselves better at thinking, remembering or running. We are born with a mix of talents and traits determined partly by biology and partly by history. We live within these limits and the decisions we make, driven by the concatenation of impulses in our brains which filter all the information impinging on us, and draw out the pattern of our history. We can never escape who we are.

Testing Times

Perhaps the greatest evidence that at our core a fractal pattern is at work comes from research done on identical twins that have been separated at birth. At first glance this evidence seems to suggest that we are simply a product of our genetics, so many are the similarities between twins separated and later reunited.

On 19 August 1939 twin boys were born in the Memorial Hospital in Piqua, Ohio. Although they were born slightly premature both boys were healthy at birth, but sadly their biological parents did not want to raise them and so the twins were put up for adoption. In the event a different family adopted each one.

Both boys were named James by their new parents, though each preferred being called Jim. From an early age they both knew that they were a twin but, to ease the burden of separation, their parents had been told that the other sibling died at birth. For thirty-seven years each twin thought he was the only survivor. Lucille Lewis had found out by accident that she and her husband

Jess had not been told the whole truth, when she returned to the probate court to complete the adoption of her baby boy. When she told the clerk of the court that she had named her boy James Edward he exclaimed: 'You can't do that. They named the other little boy James.'[22]

By whatever means the news that he had a brother filtered back to James Lewis, and as the years passed he became curious about his sibling. At first he was reluctant to try and find out about him, but eventually curiosity got the better of him and he started to search in earnest. In 1976 he found his long-lost twin. The pair lived less than 100 miles from each other in western Ohio. They met for the first time in thirty-seven years at the home of Jim Springer in Dayton, Ohio.

When they finally got together they were astonished at how alike they were. The similarities seemed to run so deep that sometimes it was hard to pick out what was different. They started with the fact that both men had been married twice, first to girls called Linda, whom they both later divorced for women called Betty. Jim Lewis had named his first son James Alan; Jim Springer had named his son James Allan. Both had owned a dog as a boy and called it Toy. Both men had taken part-time jobs as a deputy-sheriff. Both had been employed by McDonald's and both had been petrol pump attendants. Both spent their holidays at the same beach near St Petersburg in Florida, a stretch of sand only 300 yards long. Both drove to their vacation destination, all the way from Ohio. They both drove the same kind of car, a Chevrolet. Both bit their fingernails right down to the quick. Both drank Miller Lite and chain-smoked Salems. Both had basement workshops and worked in wood. Jim Lewis liked to make miniature picnic tables and Jim Springer preferred miniature rocking chairs. Both had built white benches around the trunk of a tree in their gardens. Both were fans of stock-car racing but actively disliked baseball.[23] Their health histories were very similar too. Both had suffered heart-attack scares and their blood pressure was identically high. The same eye in each one was 'lazy' and they both suffered with haemorrhoids.

The fact that they met for the first time after thirty-seven years of separation won them a lot of media attention and they came to the notice of Tom Bouchard, a psychologist from the University of Minnesota. Bouchard got in touch and called them in for a battery of psychological tests to see how alike they really were. It was uncanny. The scores for those personality traits that were considered measurable – such as sociability, flexibility, tolerance, conformity and self-control – were spookily similar. On some of the tests there was less difference between the twins than was usually found when the same person is tested twice.

The publicity surrounding the Jims – they even appeared on the Johnny Carson show – brought more sets of separated twins forth. Soon Tom Bouchard was regularly testing siblings who had only recently found out that they were one of a twin. Now the psychology department at the University of Minnesota has become one of the world's leading centres of twin research. It alone follows the lives of 8,000 twin pairs, most of them fraternal pairs. The Virginia Commonwealth University monitors the lives of 15,000 twin pairs and their spouses and siblings. And separated twins have become one of the most closely studied groups of people on the planet.

Identical Implications

Few of the subsequent twins were as exceptional as the Jims, who were such good subjects because they were separated so early. Even so studies revealing a wealth of similarities between twins have been repeated time and again.

Identical, or monozygotic, twins are special for many reasons. For a start they are rare: for every 1000 children born, only 3.5 are identical twins. More importantly they give us the chance to get a glimpse of what matters about people when it comes to working out who they are. Twins invite curiosity because of what they imply about the rest of us. We like to think of ourselves as unique, the remarkable product of our own gumption and genes. However, identical twins imply that we are a lot less special than that. They also invite curiosity because they let us test out some of our ideas about genes and environment. They let us see the relative strengths of traits and whether they break upon the hard rock of the wider world or flow to adapt. Twins are a living laboratory.

Unsurprisingly the studies of separated twins have been seized upon by those who believe that nature is everything. Those who claim that we are the sum total of our genetic information and it is this that drives the course we take in life. The fact that many twins are so similar is all the evidence they need to claim that all traits are heritable. Our IQ (whatever that is), our propensity for criminality, our attitudes to life and our willingness to pay taxes are all stored in our genes. Life is just about playing the tape.[24]

Aided by the intensive study of twins many psychologists are coming round to the view that we can make a difference to who we are, but only within the broad limits set by both nature and nurture. Set against this we have the evidence culled from the astonishing progress of molecular biology. This has led to claims by many scientists that they have found a gene for alcoholism, schizophrenia or homosexuality. Often these controversial claims are quietly withdrawn a short while after they are made.[25]

The advocates of gene-determined point of view have seized on the results of the twin studies and say that here is evidence that genes are all that matter. Nature, not nurture, is everything. Despite the fact that almost identical individuals have been brought up in different places, they turn out to be remarkably similar. The reason for this, they claim, is what they share: their genes.

Often the evidence is persuasive, almost incontrovertible. Consider the case of Amy and Beth, twin girls who were put up for adoption in New York in the late 1960s. The sisters were fair-skinned blondes with oval faces, blue-grey eyes and slightly upturned noses. Both were placed in Jewish homes in New York State in which there were siblings of similar ages. In many other respects the homes in which the little girls found themselves were very different.

Amy's family was not poor but never had a lot of money to spare. Her adoptive mother was overweight and socially awkward, her personality was flat and she had low self-esteem. She felt threatened by her newly acquired pretty daughter and extended little compassion to her. By contrast Beth's family was well off. Her adoptive mother was sophisticated, poised and graceful, and she doted on her new daughter. Psychological reports on the family noted the efforts made to include Beth and the praise and encouragement given to her. Beth's mother even dyed her hair the same colour as Beth's to minimize any differences.

Amy's mother always saw her as a problem child, a stubborn and wilful outsider. Perhaps unsurprisingly Amy started having problems at an early stage and seemed to grow into them as she aged. She was a tense and demanding child who could not stand to be left alone. She was plagued by nightmares and was scared by the world. By the age of ten she was suffering from a serious learning disorder and sought refuge in role-playing, hypochondria and gender confusion. Shockingly Beth followed exactly the same pattern. Her supportive home seemed to make no difference. She went through the same needy infant phases and at age ten ended up with the same mental problems. If anything the artificiality of her personality was more pronounced than Amy's.[26]

Genes certainly seem to rule in this case because both

a good home and a bad one turned out almost the same child.

Immovable, Irresistible

In fact the wealth of interest in twins is showing that, as is so often the case, the truth is more complicated than hard-left and hard-right politicians would want it to be. We are neither entirely determined by genes nor are we solely moulded by the world: we are both.

Many of the similarities shared by the twins can be put down to factors other than genes. Jim Lewis called his first son James Alan and Jim Springer called his James Allan, but perhaps all that reflects is parental pride and fashion. The women who adopted the little boys were not twins, nor were they related, and yet they managed to choose the same name for a new child. The names of children are not chosen randomly. The fashions, celebrities and events current at the time of the child's conception and birth influence them. For five years running now Jack has been the most popular name for a boy in the UK. For girls Chloe has been the top name for the last three years. Breakfast television presenters Richard Madeley and Judy Finnegan are thought to be largely responsible for this, because they constantly refer to their children who have these names. Few of the names in today's top fifty were popular in 1899, with the exception of Edward, George, Lucy and Emma. The reason that most of the names in the top fifty are popular is because of television.

The same could be true of some of their other similarities. They could be the result of coincidence. The fact that Jim Lewis and Jim Springer lived barely 100 miles away from each other in North America can probably explain a lot of the similarities they share. A liking for Salems and Miller Lite might again be a regional quirk. We like to think of ourselves as independent iconoclasts who are not subject to the whims of fashion but we are more alike than we care to admit. Little work has been done on the similarities twins have with people they do not share genes with. The results might be instructive. It doesn't have to be genes every time.

There are surprisingly few studies of identical twins that have been separated at birth – a few hundred cases at most.[27] Of these, few are as ideal as psychologists would want. Often the twins have not been separated from the moment of birth. Many spent years being raised alongside their twin before they were separated.

There is no doubt that genes do play a significant part in fashioning us. The question remains as to exactly how much. It is clear that they do not determine everything. It is hard to see how there could be a gene that tells people to name their dog Toy or their cat Tiger. Genes are not that specific. The instructions written in genes are carried out in a diffuse uncertain way. The influence of a gene is mediated by proteins and amino acids, which then try to force an often unwilling body into doing something. Often chemical signals from other genes are telling it to do something very different. This is not about sending reinforcements for the advance; this is sending money for the dance.

There are, however, some areas where genes do show through. Huntington's disease leads to gradual neural breakdown and death in early middle age. It has been linked to a gene on the tip of chromosome 4. With other conditions the case is not so clear cut. Some common cancers have strong genetic components. The mutated form of the BRCA1 gene has been linked to as many as half of all inherited breast cancers and many cases of ovarian cancer. With other cancers the influence of environment cannot be discounted. Women with another common genetic mutation who smoke quadruple their chance of developing breast cancer.

Height too has a strong heritable component. The studies on twins have shown a correlation of 0.90 for the heritability of height. A correlation of 0.0 means no influence and 1.0 means total conformity. This does not mean that if someone is six feet tall everything bar the last seven inches is down to genes. All it means is that 90 per cent of the part by which that person differs from the norm is due to genes. So if at six feet tall this person is ten inches taller than the average then nine of those inches will be due to genes and one will be thanks to the environment.[28]

However, even this is not clear cut. The average height of the population in many developed nations has been gradually creeping up for decades. People are taller than they have ever been. No one would take this to mean that height-increasing genes are proliferating. All it shows that a better environment, diet in the main, is helping us all grow.

With height the two forces of nature and nurture work together, but so far biologists are at a loss to explain exactly how. For traits like height and weight which do have a measurable component the situation is still very confused. The children of many Asian immigrants to the US tend to be taller than their parents are and this is probably due to the change in diet. At the same time it is hard to say how tall their parents would have been if they had been born in America. This confounding factor makes it hard to say exactly how much variance is thanks to diet.

Biologists find it very difficult to see how genes exactly influence relatively easy to measure traits such as height and weight. The result is always a meshing between environment and the individual.

Who Do You Think You Are?

The studies of twins have uncovered a paradox that needs explaining. Identical twins that are raised separately seem to be more alike than identical twins raised together. This flies in the face of claims that genes are the determining factor. Surely in a similar environment genes would play out in the same way, producing carbon-copy kiddies? Yet many of the studies indicate that the opposite is true.

The finding holds true on measurable tests of personality traits such as sociability, flexibility, tolerance, conformity and self-control, and on slightly more questionable traits such as IQ. Identical twins that have never set eyes on each other are more similar than those that are brought up together.[29] This finding is especially baffling in light of the fact that many other twin studies[30] have shown that the home in which a child is raised has little or no bearing on the way that a child turns out. This would seem to suggest that genes are the determining factor, but as other work has shown that doesn't seem to be the case either.

Psychologists think that this paradox can only be explained by considering how nature and nurture are working together. What seems to matter more than anything else is the time we spend alone. As far as developmental psychologists are concerned childhood is a time divided into shared and unshared experiences. Going to the same school as your siblings is a shared experience. Choosing to play in the orchestra rather than play rugby is a non-shared experience. Going on a family trip to the library is a shared experience. Choosing and reading books about steam engines rather than *Thomas the Tank Engine* is a non-shared experience.

Some child psychologists believe that genes do start us on the road to our adult personality and reduce the number of choices we think we have at any point. It is during these times that optimists start to see a bottle that is half-full and pessimists one that is half-empty.

Our genes play a significant part in deciding what we find interesting and how we interpret that but at all times we have the freedom to change our mind or look at something a different way. We are not entirely free because our genes do limit the choices that occur to us. Environment can obviously make a difference here. Bright toy-filled daycare centres are obviously better places for children to have formative experiences than dank cellars. The opportunities for play are much greater and the choices with which a child is faced more extensive than in a bare forbidding room. Wherever these moulding events take place it is with these non-shared experiences that the tone of a life is set. The non-shared world is the one we face alone. It is where we make ourselves and it is here that genes and environment mesh to produce something that is greater than both: You.

This helps explain why Amy and Beth went through the same mental problems even though their families were very different. Their genes started to colour the world and cut down the choices that they could make. Soon they found it hard to see the world in any other light.

It also explains why identical twins raised apart are more alike than those raised together. The twins who know nothing of each other as they grow follow their own way, and genes and environment conspire to make them look at the world in the same way. The genes they share make them experience the world in the same way and reinforce their similarities.

Twins raised together are typically treated as a single entity by their parents; they get to spend far less time alone than those raised by foster parents. As they get older they do spend more time by themselves and the evidence shows that they get more alike as they age.[31]

It is perhaps the fractal repeating nature of the brain

that drives this too. If the brain does exhibit the symptoms of Universality then that fractal pattern will be how the past is being carried forward. A fractal pattern is evidence that something is being preserved while many other things change. It is the essence of memory. The presence of a fractal is also a sign that long-range correlations are at work. Again this is how the history of a system manages to influence its future development. Small changes that took place a long time ago have a very real bearing on the way a system develops. In the same way those early formative experiences help define the view a child takes of the world and the choices that it thinks are open to it.

One of the uncomfortable implications of this work, and of Universality, is that you are not who you think you are. There is much of you bubbling away beneath your conscious experience and it can make decisions for you and determine how you approach the world. This implies that you are not wholly in charge of your life; Universality is. Fate might well be fractal.

Chapter Seven

Just U

We shall not cease from exploration and the end of all our exploring will be to arrive where we started and know the place for the first time.

Little Gidding, T. S. Eliot

The world was once all miracle. Then everything started to be explained. Everything will be explained in time. It's just a matter of waiting.

Earthly Powers, Anthony Burgess

Wonder World

We know roughly how big the universe is. We're pretty sure it is expanding rather than contracting. And we're still debating what exactly happened during the Big Bang. What we do know for sure is that Universality permeates every part of all that there is. On every scale from the titanic to the tiny the same dynamic is at work ordering and arranging.

If you could take in the sweeping vista stretching

between the farthest reaches of the universe and the smallest corners of the cells in your body you would see Universality at work anywhere and everywhere you looked. The dynamic that Universality explains has arranged the clumped clusters of galaxies and stars, the building blocks of the larger Universe, across billions of light years and aeons of time. It is evident across the almost unimaginable distances that separate one from the other. It has set the spiral splash of the Milky Way and the foggy disc of stars and stellar matter at its middle moving and got it arranged to its own stately rhythm.

Down here on earth Universality is just as busy. No matter where you rest your gaze, on the spread of clouds in the atmosphere or the shape and sharpness of the mountain ranges and coasts, if you take in the nomadic herds of wildebeest and flocks of flamingos, even the messy spreading masses of our cities, all are being arranged by the principle of Universality. The blood vessels and bronchial tubes within many of the organisms on earth and the distribution of genes within our DNA follow the same fractal pattern.

Moreover, as this book has tried to show, it does not stop there. Stock market collapses, heartbeats, athletic performances, history, even your very life might be driven by it.

The rise of Universality is an outgrowth of the intellectual revolution started by the mathematics of chaos. This seismic cerebral shift has galvanized both science and scientists and has stopped scientific disciplines shattering into narrow specialisms. Instead it has got people talking across boundaries that used to keep people penned within. It has shown the benefits of interdisciplinary research and how useful it can be to combine insights from many fields.

The rise of theories that explain complex systems has been taken as hailing a change in the way we study and understand the world. As James Gleick says in his justly popular book about the history of chaos theory: 'Where Chaos begins, classical science stops.'[1] Science is having to leave the lab behind and turn to less certain ways of describing the world and what is happening. This is no longer about inclines and accelerations, it is about inclinations and arguments. It is not photo-realism but impressionism.

In a profound change science has started to tackle the mundane, the everyday events and occurrences that season human life. With the tools and techniques emerging from chaos mathematics, complexity theory and Universality, ordinary life is being subjected to the kind of analysis usually reserved for subatomic particles or life-threatening diseases. Science is coming home and Universality is leading it there.

With chaos theory came wonder. Its breathtaking insights showed that there was structure to the world and its myriad events even if it was sometimes hard to pull free. Although chaos mathematics may have laid bare the world's structure it has done little to advance explanations. It showed that weather systems could be inherently unpredictable but did little to say how or why.

With complexity theory and Universality comes understanding and acceptance. The dynamics, the hows and whys, are being made clear and explanations are forthcoming. Adopting these explanations has meant abandoning notions of control or predictability but this is a small price to pay for the profound insight and ease we have gained. Now we can reflect the world in all its splendour. Universality implies that the world is not capricious, cruel or out to get you. We can relax knowing that the wellsprings of so much in our lives are readily understandable. It may be unpredictable but not so much so that you can do nothing about it. Our brains seem to be very good examples of a system existing at a critical point. They are very good at logging events and recognizing patterns, even those that are hard to pick out because of their near unpredictability.

MIT psychologist Gary Klein believes that this intimate understanding makes itself felt in hunches, those gut feelings that something is about to happen or go wrong. For the last fourteen years he has been collecting evidence for hunches. His work suggests that people who are intimately familiar with a subject or a job can make accurate predictions about what will happen next even if they don't know why they know. Klein has hundreds of stories demonstrating that we can make life-saving decisions on the basis of incomplete information. Perhaps we are running scenarios in our heads and exploiting the natural tendency of our brain to follow this rhythm.

Universality is giving us our humanity back. No longer do we have to accept that mystical forces are at work forcing our feet onto the marks written by divine design. Now we see that the dynamics within us are writ large and reflected all around us. This is where we belong and all the explanations are right here. The world is godless but not loveless or charmless.

Michael Barnsley's warning about fractals[2] applies just as well to Universality – there is a danger in knowing about this subject. Knowing how fractals are formed, their dynamics alerts you to their presence. And seeing them everywhere changes the way that you regard the world. Childhood beliefs and some of the remaining mystery, superstition and religion has been stripped away and replaced with a different way of viewing the world.

Writing this book has sharpened my senses to hints of Universality and emergent orders. Everywhere I go now I look to see if it is at work frustrating people, forming phantom traffic jams or making the grass on my lawn grow. I've taken a look around and, as far as I can see, this stuff is everywhere.

There are those who think the world of Universality. Former US Vice President Al Gore has written glowingly of Per Bak's work on self-organizing criticality, saying: 'One reason I am drawn to this theory is that it helped me understand change in my own life.'[3] Many people may not go as far as Mr Gore but knowing about Universality can engender a sense of peace with the world. It can help you relax by making you realize the limits of your knowledge and the changes you can make.

Even though Universality is ubiquitous it is not the only explanation we need, and we should not overestimate its usefulness. It would be very easy to do so given the grand sweeping name it has acquired, but we should be wary of overselling it, seeing its influence everywhere, and claiming that it is the only explanation we need to make sense of the world. It isn't the be all and end all.

It is worth reiterating that work on Universality is at a very early stage. The paper by Bak, Tang and Wiesenfeld on self-organizing criticality published in 1987 kicked off an avalanche of work on the whole subject of critical points. Some of it is good, some bad. The appeal of that idea and Universality meant that many scientists were flushed with enthusiasm and claims are now being made for the dynamic that may not be borne out.

To begin with there are lots of things that can be explained without recourse to Universality at all. It only applies to a small slice of all the possible states some systems with certain characteristics can assume. Above and below this point other principles that have been established much longer can explain everything that is happening. Classical thermodynamics does all that you need. Similarly the distribution of many other phenomena are not brought about by Universality. If you plot the heights and weights of any random sample of people you will get a bell curve or normal distribution. There will be some tall people and some small, but the majority will fall in the middle. There is no need to invoke Universality here. There are a lot of theories that make the world meaningful.

All kinds of claims are being made for Universality largely because it chimes with the prevailing feeling of our times. The network has become the metaphor of our age. We are gradually starting to realize that nothing exists in isolation. Everything is interconnected in some way to everything else. The rise of interdisciplinary sciences such as chaos and complexity theory has helped bring about this change in attitude.

We no longer see species as individuals; instead we try and take into account their relationships with other organisms. Boundaries are coming down all over the place. Within your gut live hundreds of millions of microorganisms – mainly non-lethal strains of the bacterium *E. coli* – that exist in symbiosis with us. Millions of others exist on your skin. We rarely pay these bugs attention, and yet without them we would not be able to survive. We need them as much as they need us.

Climate research has shown the profound effect that we busy humans can have on the planet, and the destruction of the ozone layer has brought home to us the consequences of our actions. We do not exist in isolation. Crashes on the stock market rattle around the world and affect us all. It is perhaps no surprise that the Internet is proving so popular of late. It encapsulates the feelings we have about us all being interconnected. We are now realizing that we are all in this together.

Every age adopts the ideas that match the prevailing world-view and those ideas are outgrowths of popular beliefs. There is a lot of interaction between the two. The way scientists think about their subject and the uses that society puts them to are interdependent.

History shows us the value of being careful. In the late seventeenth and early eighteenth centuries the success of Newtonian physics kindled a movement that tried to apply his ideas to all walks of life. Charmed by the simplicity of the ideas in the *Principia*, many English and continental philosophers looked for similar laws in the terrestrial world. The Newtonian conception of the heavens as an ordered, highly regular system prompted theologians to search for the laws of bodily, rather than celestial, motion.

In 1699 a mathematician called John Craig tried to do for theology in his book *Mathematical Principles of Christian Theology* what Newton had done for science. Not only was the title of Craig's book modelled on Newton's *Mathematical Principles of Natural Philosophy*, but in it he gives the three simple laws which he claimed could be used to explain much of Christian theology. Craig viewed belief in much the same way that Newton saw inertia. A thrown rock will gradually have its inertia overcome by gravity and Craig thought that beliefs were subject to the same diminution. As time went on they would gradually fade. Using this starting point Craig worked out that the Second Coming would have to happen before 3150. Beyond this date all belief in Christ would be dissipated. The inertia of belief could carry it no further. Prior to this date there should be at least one person around that still believed in Christ to ensure that the Second Coming was not a complete non-event.

Few other philosophers were quite as literal as Craig but many men did try to apply Newtonian ideas to the mortal world. In 1706 influential theologian Samuel Clarke – a chaplain to the Bishop of Norwich and later Queen Anne – published a book entitled *A Discourse*

Concerning the Unchangeable Obligations of Natural Religion. In this he tried to derive the existence of God using theorems and propositions just like Newton did.

Newton had a huge impact on both science and society. His vision of the heavens operating according to the dictates of immutable laws chimed with contemporary views about the working of the creator and an orderly regular universe. Newton's ideas were misused in his day largely because it seemed to back up the religious beliefs of the time.

Steam-Driven Science

It should be obvious that every age can only adopt the ideas and metaphors available to it. The ideas we can form are limited by the completeness of our knowledge. At the same time society puts those scientific theories to all manner of uses that were not initially intended for them.

When Charles Darwin published *The Origin of Species* in 1859 his ideas must have come as a great shock to many Victorians – particularly its claim that men and monkeys were cousins. Once Victorians got over the shocking message of Darwinism they started to like it because it seemed to excuse the injustice they saw all around them. They lived in a world where only the strong survived and the weakest went to the wall and Darwin gave them the justification that helped them sleep at night. Karl Marx criticized Darwin for his conception of nature that turned it into yet another market ripe for exploitation. By the same token this turned the social world into a place where rampant competition was natural.

Instead of worrying that they were mistreating their fellow men by paying them low wages the industrialists could argue that the poor were not fit to succeed or in some cases survive. As a result they deserved whatever they got. The factory and mill owners by contrast were confirmed in their superiority because they were in charge.

This interplay of ideas has been repeated many times. Prevailing scientific theories have been co-opted by other fields and the population at large. There is a definite whiff of the steam engine in many of Freud's ideas. It is hard to read his essays on psychoanalysis, with their talk of energy being channelled, contained or breaking out if the pressure gets too much, without thinking of a steam engine. Unconsciously he may have phrased it that way because that was one of the dominant metaphors of his day. As a doctor, though he received his Ph.D. for studying the gonads of eels, it was probably one of the few pieces of technology he was familiar with.

The same thing happened in the 1960s when computers started to become popular and useful. They might still have been as big as a car but at least governments and businesses were starting to use them in large numbers. Many psychologists saw the early success of computers and assumed that the brain worked the same way. All they talked about was getting the right input and output. No one worried about what happened in the middle because it was assumed that there was just a general processor of some sort inside. This led to nearly forty wasted years as psychologists used the wrong metaphor to understand their central object of study. Now our understanding of the brain shows just how wrong that view was. Now we know that the brain is nothing like a computer.

Now more organic and networked metaphors are starting to take over. James Lovelock's conception of Gaia – a planetwide interdependent ecosystem – has become something of a new orthodoxy. It is hard to see how it could be otherwise. We appreciate our connectedness and feel the weight of our social ties at all times. Economies, businesses and social programmes are described using biological metaphors.

Nonetheless, we should be just as wary of assuming that the way we think of the world is the right one. The experience we have had of being wrong for so many centuries should make us cautious about claiming too much for ourselves now. We should realize that all certainty is temporary and is rooted in the society that gives it life.

We should also realize that many ideas have been around for a long time but only become popular when they fit so well with a particular view of the world. Many of the ideas that we now know as chaos theory were around for decades. In 1941 Joseph Schillinger wrote about many of the ideas that Benoit Mandelbrot would later become associated with. Lewis Richardson did a lot of work that inspired Mandelbrot. Even Mandelbrot himself worked tirelessly for years on his ideas before other scientists and, later still, the public at large became interested.

For these reasons we should be slightly suspicious of the wealth of claims made for Universality. Much of its appeal is due to the fact that it fits in with a popular view of the world that is now starting to emerge. Its popularity is also helped by long-established Christian traditions that stress ultimate plans and all-encompassing principles. The notion of a theory that unifies everything is the express aim of many physicists and they are motivated more by theology than they would probably care to admit. It leads many physicists to search for God in the equations of their science. However, Universality should not be taken in these terms. It is not the be all and end all.

Even some of the phenomena that Universality seems to make a good stab at explaining are turning out to be the result of something else. Zipf's law which describes the distribution of words in a huge variety of texts can be mimicked using monkeys bashing on typewriters.[4] The language that they produce may be rather wordy because they are more likely to hit a letter than the space bar to separate words but it shares many of the properties that Zipf claimed to see in written languages. Mandelbrot considers Zipf's law to be little more than a statistical numbers game that doesn't really prove anything,[5] or at least does not prove what people would like it to.

U or Non-U

For this reason it is worth being sceptical about just how much Universality can explain. Leo Kadanoff and Ken Wilson did the pioneering work on Universality in the late 1960s and early 1970s but it is only now that the scope of the idea is being fully explored. In the last decade sweeping claims have been made for the theory that are still being examined.

The amount of research done on Universality has increased lately, largely because it is easier to learn programming than it ever has been and now computer power is cheap. Instead of using a mainframe and asking a qualified colleague to do the job, many physicists are using desktop computers to run their experiments. Building a simulation has become much more straightforward and, because of the simplicity of the dynamics at the heart of Universality, has made it easy to achieve convincing results. This ease is seductive and should always be balanced with experimental work. The disparity between the results achieved by Per Bak, Chao Tang and Kurt Wiesenfeld on a computer and the trouble other researchers have had recreating those results in the real world is a case in point.

The toy models that Bak, Chao and Wiesenfeld created seemed to show that complex systems could drive themselves to a stable critical state – hence the name of self-organizing criticality. In their simulations the trio created a computer model of a sandpile. However a raft of attempts to recreate these results in the real world using real sand, rice, mud and other media have largely failed. Any evidence for the distribution of results that do back up self-organizing criticality in these materials persist over very short time spans.

Experimentalists have found it very hard to persuade real-world systems to achieve a state that Bak and his co-workers claim they should not be able to avoid. But this does not mean that the dynamics and properties Bak and his colleagues have laid bare are illusions. Kevin O'Brien from the University of Illinois and James Sethna from Cornell have produced exactly the same scaling distribution of events seen in self-organizing criticality without the dynamics Bak claimed were essential. They have shown that even when $1/f$ noise is present it does not always have to be caused by self-organizing criticality.[6]

This is not just a problem for Bak and his colleagues. At one time or another the whole field of Universality has been criticized. Some people think, frankly, that all of it is nonsense. They say that there are much better explanations for $1/f$ noise and many of the short-term fractals that can be seen in the larger world. They see no need for invoking a theory with the scope of Universality.

The biggest problem that critics of Universality have with the theory is the evidence they marshalled to back it up. Many of the systems claimed to show dynamics reminiscent of those seen at a critical point persist only over a couple of orders of magnitude. In some of the systems under study this is unavoidable because of the limited size of the system in question. Measurements of heartbeat dynamics cannot extend lower than the protein chains in the muscle cells and no further than the whole organ. For many proponents of Universality this is enough, but critics say it is claiming too much. They decry attempts to establish a Universal principle on very short-term evidence and a lot of hand-waving about the way that the relationships would hold up if only organs or the systems under study were bigger.

Critics also say that there are better ways mathematically to capture the distribution of events in systems that have been claimed for Universality. Didier Sornette from the University of Nice and his colleagues claim that this data fits a distribution known as a stretched exponential far better than they do any power laws. A stretched exponential is one in which the exponent is smaller than one. Exponential growth involves something doubling at every step but the growth of a stretched exponential tails off. With both power laws and stretched exponentials large events are rare.[7]

Sornette claims that stretched exponentials can be used to account for the distribution of radio and light emissions from galaxies, the sizes of US oilfield reserves, UN country sizes, daily Forex US\$–DM and Ffr–DM price variations, the distribution of biological extinction events, the distribution of the largest earthquakes in the world from 1977 to 1992, citations of the 1120 most-cited physicists from 1981 to June 1997, and temperature data for over 400,000 years from Vostok near the south pole.[8] It reads like a list of all the systems that Universality is claiming for itself. As yet Sornette has not come up with a mechanism that can produce stretched exponentials. In contrast the mechanism behind fractals and power laws is well established. Its explanatory power is being demonstrated on a regular basis.

It is true that few scientists working on Universality would want to claim that the systems they are studying produce geometrically perfect fractals. The important point is not absolute self-similarity but statistical resemblance. Are they sufficiently alike? At different scales coastlines and fractal trees in plants and lungs do not have to be perfect matches. They just have to be close enough. This leeway explains the divergence from power laws and leaves many scientists satisfied that what they are dealing with is a system at a critical point.

A further blow was dealt to the credibility of Universality in late April 2000. Physicists have come to understand the dynamics of the critical point using an idealized simulation called an Ising model. It was first developed by Ernst Ising in 1926 as part of his Ph.D. dissertation.

Ising's original model imagined a line of magnet-like particles that could assume one of two positions, up or down. The position of each magnet influenced that of its neighbours. It has obvious affinities with the canonical critical system – a hot bar magnet. In 1946 Lars Onsager expanded the Ising model into two dimensions and extended its applicability to many more systems, typically those that undergo phase transitions. Over 8000 papers using the Ising model were published between 1969 and 1997. It has been key to the development of the principle of Universality. Later work has extended this model into the third dimension but not with the same degree of accuracy as that of the one- and two-dimensional versions. Many physicists have blithely assumed that it will apply in three dimensions and extend its reach even further.

Now Sorin Istrail from the Sandia National Laboratory in Albuquerque has shown that the Ising model cannot be extended to three dimensions. Istrail has proved that the 3-D Ising model falls into a class of mathematical problems that are computationally intractable. It cannot be done.[9]

This does not mean that the Universality no longer applies. Many of the scientists working in this field no longer rely on the Ising model to capture the dynamics of what they are studying. Cheap computer power has opened up other avenues of research. Furthermore much of the work done on identifying systems displaying Universality uses real data rather than the approximations of the model.

There are problems with extending Universality to the human sphere and using it to understand social change largely because of the imprecise nature of the quantities available to measure. Winn Farrell might claim that Universality is at work in the buying and selling of records but there are no hard and fast figures that capture what is meant by one pop star being more or less popular. Most of the measures that do exist are problematic. We could record sales as a guide for popularity but we know that there are all kinds of ways to manipulate the numbers sold. Promotions, cost cutting and marketing can all boost figures artificially and thereby dilute the explanatory power of any model let alone that of Universality. Even something like the level of the Standard & Poor's 500 index is more complex than it first appears. The index is measured in dollars the value of which fluctuates on a daily basis and complicates any measure.

Critics of Universality will take a lot of the musing about fate and fractals to be at best science in the service of speculation and at worst mere supposition. Certainly the work on brains, fate and Universality is at a very early stage, but even this far the evidence is suggestive. The scientists doing the work believe there is value in asking the questions and playing with the ideas, but it must be said that as yet all applications of Universality to these areas should be taken with a pinch of salt.

This problem confronts, not just fans of Universality, but every physicist, biologist and chaotician who studies complex systems. Physics has long left behind the certainties of ultimate reductionism. Instead statistical theories, such as that for quantum mechanics, are becoming the norm. It is impossible to describe these systems mathematically because they are so complex. Instead we are left with approximate models and probabilities rather than certainties. In such a situation physicists rely on models to capture what they are studying. These simulations are constantly being refined to better reflect reality.

Biology rarely deals with hard facts. Much of biology is naturalism – finding out how an organism lives its life. In such a description contingencies and accidents play a huge role; there are no certainties. While the workings of a body are subject to the principles of chemistry and physics it takes more than a grasp of those fields to grasp what matters to a fox or a human. Even if computers become much more powerful than they are now we are never going to be in a position of being certain about many of the complexities of the natural world. Uncertainty, probabilities and contingencies are the order of the day. We are going to have to get used to it.

Where the data is available, however, there is no doubt that a common dynamic does seem to be pulsing within a vast range of systems. It is at work everywhere from the smallest to the greatest scales, even with something like the weather. The global weather system is often seen as the canonical chaotic system. Its capriciousness makes it impossible to predict what it is going to do for more than a couple of days hence. Often the only thing you can say about what will happen tomorrow is that the weather will be like it was today but slightly different.

Yet it seems to be the case that there are long-term correlations at work within global weather patterns. Armin Bunde from the University of Giessen in Germany has found that there are temperature relationships that persist across centuries. Bunde took data from fourteen randomly chosen sites as far apart as Vancouver, Melbourne and Prague and tried to find out if there was any relationship between the weather in one part of the world and all the others. He asked, if it is hot and sunny today, what are the chances that it will be the same tomorrow, next week or next year? As expected the further away from the starting date he got the less similar was the weather, but it did not change as fast as he expected. It did not decay exponentially but according to a power law. Like all statistical laws this cannot be used to predict temperature on any particular day but it does have its uses. Bunde expects to use the results to calibrate computer models of global climate to ensure that they are accurate.[10]

Attitude Adjuster

These theories were supposed to take science out of its bright shiny laboratories and make it engage the real world. Scientists were supposed to start getting their hands dirty and bring science home. It was supposed to be out with the mechanistic, reductionistic and atomistic view of the world and in with something more holistic. Even the rise of molecular biology could be condemned in the same way because it reduces everyone to a blind collection of genes instead of considering them as a whole.

Universality implies a return to good old classical physics. The central model – a system at a critical point – is described using statistical mechanics, one of the oldest branches of natural philosophy. You don't get much more classical than work on gases, pressures and temperatures. This is not about throwing over old ways of thinking; it is a return to basics.

This implies that to use chaos and complexity theory as a stick to beat physicists is misguided. In the end it turns out that a huge array of complex systems can be reduced to a simple dynamical model – not to their constituent elements but stripped of much of their detail. This is not the reductionism that Pierre Laplace thought was possible but it is a long way from the cosy holistic viewpoint that many people assume goes hand in hand with theories of complexity.

It is as well that Universality can reduce all these systems to simple models and that the details are irrelevant. If this were not the case we would have to study every system individually and work out the engines of change in each one. With Universality the world blooms. It opens right up and we can gain a profound glimpse of the workings of the world we live in. This is one of the more important insights to take away from Universality: the impossibility of predicting what will happen in a huge variety of situations. It shows that despite these caveats and criticisms Universality does have a profound message for anyone who cares to hear it.

While Universality is just one among many theories needed to understand the world it demands more attention than most because of what it claims to be able to explain. It applies in those situations that until now have defeated almost all analysis, events that have confounded human understanding for millennia. It can be used to understand earthquakes, weather patterns, epidemics, and perhaps even the roll of events that we call history. It seems to be the defining feature of all complex systems.

U are Everything

Universality implies that our desire to know everything is doomed to failure. It shows that many of the complex systems that make up and punctuate our lives are essentially unpredictable. We should not strive to understand what will forever remain unpredictable. Change is part of the deal, so we would do well to get used to it. If you are looking for certainties you have come to the wrong existence.

This acceptance should not engender a feeling of defeat. We should embrace and enjoy the unpredictability of the world. Universality implies that experience is a very good guide because, although change can propagate rapidly and events are unpredictable, they are a result of contingency and history. There is always a cause, no matter how far back you have to go. Universality emphasizes the interconnections between the elements of a system, be they the neurons in your brain or droplets of water vapour in a cloud. What happens today is determined in part by the events that have gone before. Possibly long before. Correlations can persist over arbitrarily long distances, be they spatial or temporal.

You might even find comfort in the fact that many of the events that plague your life – traffic jams, delayed buses, rainstorms – are not meant to inconvenience you personally, they are just the workings of the world. There is no point cursing the blind workings of this system. Far better to keep cool, accept that it is the way that the world is working, and perhaps try and guess what might happen next. You are probably better at it than you think.

Although systems displaying Universality are essentially unpredictable they can be modelled accurately. In your brain you might have one of the best examples of a system at a critical point. You may be able to mirror in miniature the dynamics playing out in the wider world.

There is a growing body of evidence that shows that hunches are worth following. On a preconscious level we might be running through a simulation and getting a hint of what might happen next.[11]

Deep down we know that the world is not really chaotic and capricious. We know that it has a rhythm to it even if it is one that we cannot hum or write down. We can grasp it but not name it. If life were not predictable in some sense then at every moment we would be confronted by catastrophe. Experience would be no help at all; we would have to live every moment on our wits because anything could happen next.

If life were utter chaos, random and unfathomable we would have no drama, no tragedy, and no comedy or farce. If anything could happen next there could be no dramatic tension. We would have no reason to feel shocked, shamed or surprised by what we saw on the stage or television. There would be no room for affection and love.

The fact that we have Universality built in should be another source of comfort. The rise of science has pushed back the boundaries of what religion can claim for itself. In knocking down dogma science has not put anything in its place. It has let us feel the keen wind of agnosticism and see nothing but the cold light of the stars above. In contrast Universality shows that we are wedded to the Universe in the most liberating and intimate sense. We are intimately connected to the cosmos, but not only because all the elements that we are built from were made long ago inside stars.

At the critical point systems in the same Universality

class are identical. So if we have Universality twisting within us, forming our bronchial trees and keeping our heart hammering, then we share the dynamics of the universe. We have a place in the cosmic scheme. We are indelibly part of the whole and, at times, indistinguishable from it.

You see Universality at work every day. You can see it in the way that people spread themselves out on a crowded train. People will walk through lots of carriages just to find a spot that is not too close or far away from other people. Often they meet people coming the other way. They are not searching for just any seat; they are searching for the right seat – one that preserves social rules about how close people should get. You can also see it in the bunching of cars on the road, the spread of daisies across a field, in cloud patterns, crackles on a phone line, the wrinkles on your skin and the veins in the back of your hand. Unseen, the rhythms still ring and they encompass us all.

The processes that produce you and me are ubiquitous, from the smallest scales on earth to the spread of stars across the heavens. We lie somewhere in the middle. We are very definitely a part of this universe. We are one with it; we do not stand outside it. We are reflections of each other. We should not feel left out and alone when at our heart we hear the pulse of our belonging every moment of every day of our lives. But the price of this belonging is that we become nothing special. We are part of a self-similar pattern reflected throughout the universe. This shows that a very common dynamic can give rise to structures, like you and me, that persist in defiance of the laws of thermodynamics. Again this emphasizes our intimate link to the universe and Universality even if it removes us from the pedestal that we sometimes place ourselves upon.

It is early days for Universality. Much of the work that will firmly establish it has yet to be done. We are still in the naturalist phase of this theory. We are still mapping all the places that it turns up, rather than working out what exactly it means. So far many of the results are approximate and there are large gaps in our understanding.

Even if it is the case that Universality turns out to be only one way to characterize all those fractals and those driven dynamics, there is still comfort here. Even if Universality is not the theory that can explain these dynamics, the rhythms remain. Nothing can detract from the fact that a common cadence has been heard and all of us can listen to it if we choose. We cannot forget or deny what we now know for sure. We share a thrilling pulse with a huge part of this world, the solar system and the universe at large. We belong here. We know our place. We know our place and we are home.

Endnotes

Chapter 1. The Wonder of U

1. French (1972), 4.
2. Shumaker (1982), 22.
3. Casaubon (1659), 3.
4. Casaubon (1659), 3.
5. Casaubon (1659), 4.
6. Shumaker (1982), 27.
7. Casaubon (1659), 6.
8. Yates (1964), 1.
9. Yates (1964), 13.
10. Wertheim (1997), 85.
11. On 12 March 2000 the Roman Catholic church held a Request for Forgiveness Day on which Pope John Paul II sought forgiveness for errors made by the Church and the injustices it had committed. On that day Bruno and others who had suffered at the hands of the Church received a formal apology. Despite this the Church says there is no chance that Bruno will be rehabilitated back into the Church because it regards his teachings as contrary to Christian doctrine.

12. Thomas (1971), 268.
13. The name comes from the Greek *planetes* which means 'wanderer'.
14. Gurshtein (1997), 271.
15. Wertheim (1997), 26–27.
16. Quoted in Wertheim (1997), 10.
17. Thomas (1984), 88.
18. In Newton's pantheon of thought, science and magic were almost indistinguishable. Some have speculated that the only reason that Newton was prepared to accept that gravity could work over vast distances was because of his familiarity with magic which claimed all kinds of sympathies for itself. See Wertheim (1997) 117–118 for an exploration of this issue.
19. Aristotle (1998), 4.
20. Sartre (1948), 28.

Chapter 2. The Littoral Truth

1. Wilson (1983), 583.
2. One of the best treatments of a bar magnet at the critical point can be found in Bruce and Wallace (1989), 217.
3. Atkins (1984), 10.
4. Wilson (1983), 585.
5. Stanley *et al.* (1980), 24.
6. See Wilson (1983), 588, for a more complete list of the papers detailing the shortcomings of mean field theory.
7. Kadanoff (1999), 157.
8. Kadanoff (1999), 297–299.

9. Kadanoff (1999), 159.
10. Kadanoff (1976).
11. Kadanoff (1999), 159.
12. Kadanoff *et al.* (1967); Kadanoff (1999), 159.
13. Kadanoff (1999), 159.
14. Gleick (1988), 161.
15. Bruce and Wallace (1989), 236.
16. Kadanoff (1966).
17. Bruce and Wallace (1989), 246.
18. http://www.nobel.se/physics/laureates/1982/press.html.
19. This is subject to the limits of the system in question, of course. For the magnet this would be from a position so close that the individual miniature magnets became distinct to the point where you could take in the whole bar with a glance.
20. See Wilson, Kenneth, 1971. 'The Renormalization Group and Critical Phenomena I: Renormalization Group and the Kadanoff Scaling Picture' *Physical Review* B. Vol 4 3174. Wilson, Kenneth, 1971. 'The Renormalization Group and Critical Phenomena II: Phase Space Cell Analysis of Critical Behaviour' *Physical Review* B 4. 3184. Wilson, Kenneth, 1971. 'Renormalization Group and Strong Interactions' *Physical Review* D. Vol 3. 1818.
21. Mandelbrot (1977), 3.
22. Mandelbrot (1977), Foreword.
23. Bak (1997), 24–25.
24. Bak (1997), 25–26.
25. Mandelbrot (1977), 4.
26, Mandelbrot (1967).
27. Mandelbrot (1977), 25.
28. This is not just an abstract problem. The encyclopaedias

of different countries rarely measure the lengths of borders with neighbouring territories using the same unit scale. Often the measures can differ by as much as 20 per cent. See Mandelbrot (1977), 27.

29. Another great guide to self-similarity and power laws is Schroeder (1991).
30. Weibel (1993), 81.
31. Lam (1998), 14.
32. Bak (1997), 21, and Huberman *et al.* (1998).
33. Bak *et al.* (1987).

Chapter 3. The Midwife of Creation

1. Deamer (1997).
2. Schroeder (1991), 30–31.
3. Coveney and Highfield (1995), 106.
4. Gouyet (1996), 90.
5. Margulis and Sagan (1986), 79.
6. If this mission plan sounds science-fictional this is not surprising because the pre-planning board that dreamt it up counted among its members noted SF authors Gregory Benford, Greg Bear and Larry Niven.
7. Wolfram (1984), 419.
8. Wolfram (1984), 419.
9. Anyone interested in popular accounts of artificial life could read Steven Levy's (1992) book or my own *Virtual Organisms*. Between them they do justice to the whole gamut of ALife research.
10. Langton (1992), 84–85.
11. Langton (1992), 85.
12. Curiously one of the phenomena that Benoit Mandelbrot

worked on and piqued his interest in fractals was the self-similar nature of noise on telephone wires.

13. Langton (1992), 85–86.
14. See Strong and Ray (1975) and Ray (1976) for research on the vines.
15. Adami (1998), 274–276.
16. See Bak *et al.* (1987).
17. The paper laying out the basic ideas of self-organizing criticality was published in 1987 in *Physical Review Letters*. However, Bak (1997) is a far more approachable explanation of the central ideas of SOC.
18. See Jaeger *et al.* (1989) and Jaeger and Nagel (1992).
19. See Held *et al.* (1990).
20. Jensen (1998), 15–16.
21. See Frette *et al.* (1995).
22. What makes it difficult to see critical phenomena in the everyday, as opposed to idealized conditions on computer, is the limited reach of many real world events. Few encompass anything like enough data to be sure they are wearing a critical signature. See Chapter 4, pages 168–72 for a longer discussion of this point.
23. See Somfai *et al.* (1994).
24. See Kauffmann (1993), Chap. 12.
25. Lewin (1993), 55.
26. Jones (1993), 5.
27. Holmes (1995).
28. See Stanley (1969).
29. See Munson *et al.* (1992).
30. See Maddox (1992).
31. See Buldyrev *et al.* (1993).
32. See Freeland and Hurst (1998).

Chapter 4. The Rhythm of Life

1. Suki *et al.* (1998), 128.
2. The heart music can be found on the Internet at http://polymer.bu.edu/music.
3. See Voss and Clark (1978).
4. Gardner (1978).
5. Schroeder (1991), 107.
6. Phil Thompson's music can be found on the Internet at http://www.organised-chaos.com.
7. See http://news.bbc.co.uk/hi/english/sci/tech/newsid_128000/128906.stm.
8. Schroeder (1991), 109.
9. Schroeder (1991), 109–112.
10. Perrett *et al.* (1994).
11. See Benson and Perrett (1992).
12. See Shapir *et al.* (2000).
13. Lewin (1998), 36.
14. See West *et al.* (1997).
15. See Avnir *et al.* (1998).
16. Avnir *et al.* (1998), 39.
17. Mandelbrot (1998), 783.
18. See Hausdorf *et al.* (1997).
19. Coleridge's *Rime of the Ancient Mariner* revolves around the superstition that to kill an albatross is to invite a terrible fate. Yet in Patrick O'Brian's series of novels about the same period, which are as authentic a historial recreation as you are likely to get, sailors are portrayed as a superstitious bunch but there is no mention of albatrosses being unlucky. In several of the books the marines shoot

the albatrosses so that Stephen Maturin can indulge his dual passions for ornithology and anatomy.

20. Viswanathan is a colleague of the indefatigable Gene Stanley.
21. Stanley *et al.* (1996a), 314.
22. Holldober and Wilson (1994), 160–161.
23. Strogatz and Stewart (1993), 70.
24. Goodwin (1994), 64.
25. Merritt (1985), 6.
26. Merritt (1985), 43.
27. See Bonabeau *et al.* (1999). 4476.
28. Keitt and Stanley (1998), 258.
29. Jones (1993), 126.
30. Newman (1997), 235.
31. See Alvarez *et al.* (1980).
32. Newman (1997), 236.

Chapter 5. The Business of Complexity

1. *How Hits Happen* was the name of Farrell's (1993) book about complexity theory and how it can help make sense of the dynamics of popularity.
2. Hootie and the Blowfish are nicknames for two friends of the band.
3. Farrell (1993), 26.
4. See Petzinger (1997).
5. Farrell (1993), 20.
6. Gleick (1988), 84.
7. Mandelbrot (1977), 339.
8. Mandelbrot (1963).

9. See Chapter 4, pages 168–72 for a more in-depth discussion of this point.
10. Bak (1997), 186.
11. See Gopikrishnan *et al.* (1998).
12. Mantegna and Stanley (2000), 36.
13. For some reason *The Eudaemonic Pie* became *The Newtonian Casino* when it was published in the UK.
14. See Kephart *et al.* (1998).
15. Waldrop (1992), 23.
16. Bak (1997), 185–187.
17. Mantegna and Stanley (2000), 36.
18. Matthews (1999), 27.
19. See Amaral *et al.* (1997), 3, for a list.
20. Hart and Oulton (1995).

Chapter 6. Me and U

1. Day (1996), 5.
2. Solé *et al.* (1999), 157.
3. Janssen *et al.* (1999).
4. Diamond (1998), 211.
5. Schama, S. (1996), 6.
6. Huberman *et al.* (1998), 96.
7. Barabási *et al.* (2000), 850.
8. Savaglio and Carbone (2000), 244.
9. Ayres(1999).
10. Taylor *et al.* (1999).
11. Holmes (1998), 32.
12. Spinney (1998), 42.
13. Smith *et al.* (1996).
14. Miller (1962), 50–52.

15. See pages 81–2 for a list of critical system traits.
16. Jung (1997) and Jung *et al.* (1998).
17. Jung *et al.* (1998).
18. Kelso and Hakken (1995), 151–156.
19. Stewart (1990), 232–233.
20. Gleick (1988), 157–187.
21. Miller (1956).
22. Wright (1997), 38.
23. These lists are adapted from Wright (1997), 38–42, and Watson (1981), 9–12.
24. The majority of those who use the twin studies to excuse the status quo usually have a political axe to grind. It is unlikely that science is ever going to change the interpretations that people put on these studies.
25. Jones (1993), 237–241.
26. Wright (1997), 2–5.
27. Wright (1997), 60.
28. Wright (1997), 23.
29. Wright (1997), 120.
30. Wright (1997), 38–39.
31. Wright (1997), 120.

Chapter 7. Just U

1. Gleick (1988), 3.
2. See Barnsley (1988).
3. Bak (1997), 62.
4. Schroeder (1991), 35.
5. Yam (1995), 12.
6. See O'Brien and Weissman (1994) and Perkovic *et al.* (1995).

7. Laherrere and Sornette (1998), 527.

8. Laherrere and Sornette (1998), 527–537.

9. See Istrail (2000).

10. See Bunde (1998).

11. Spinney (1998), 42.

Bibliography

Adami, Christoph, 1998. *An Introduction to Artificial Life*. Springer, New York.

Alvarez, L. W., Alvarez, W., Asara, F., and Michel, H. V., 1980. Extraterrestrial cause for the Cretaceous–Tertiary extinction. *Science*, Vol. 208: 1095–1108.

Amaral, L. A. N., Buldyrev, S. V., Havlin, S., Maass, P., Salinger, M. A., Stanley, H. E., Stanley, M. H. R., 1997. Scaling behaviour in economics: The problem of quantifying company growth. *Physica A*, Vol. 224: 1–24.

Aristotle, 1998. *The Metaphysics*, transl. Hugh Lawson-Tancred. Penguin, London.

Atkins, Peter W., 1984. *The Second Law*. Freeman, New York.

Avnir, D., Biham, O., Lidar, D., and Malcai, O., 1998. Is the Geometry of Nature Fractal? *Science*. Vol. 279: 39–40.

Ayres, Alison, 1999. Homing in on a serial killer. *Daily Telegraph*, 'Connected' supplement. 22 April 1999, pp. 8–9.

Bak, P., 1997. *How Nature Works*. Oxford University Press. Oxford.

Bak, P., Tang, C., and Wiesenfeld, K., 1987. Self-organised criticality: An explanation of $1/f$ noise. *Physical Review Letters*, Vol. 59 No. 4: 381–384.

Barabási, A.-L., Néda, Z., Ravasz, E., Brechert, Y., and Vicsek, T., 2000. The sound of many hands clapping. *Nature*. Vol. 403: 849–850.

Barnsley, Michael, 1988. *Fractals Everywhere*. Academic Press, New York.

Benson, Philip, and Perrett, David, 1992. Face to face with the perfect image. *New Scientist*, 22 February, Vol. 133 No. 1809: 32.

Bonabeau, E., Dagorn, L., and Freon, P. 1999. Scaling in animal group-size distributions. *PNAS*, Vol. 96 No. 8: 4472–4477.

Bruce, A. D., and Wallace, D. J., 1989. Critical point phenomena: Universal physics at large length scales. In: *The New Physics*, ed. Paul W. Davies, pp. 236–267. Cambridge University Press, Cambridge.

Buldyrev, S. V., Goldberger, A. L., Havlin, S., Peng, C.-K., Stanley, H. E., Stanley, M. H. R., and Simons, M., 1993. Fractal landscapes and molecular evolution: Modeling the myosin heavy chain gene family. *Biophysical Journal*, Vol. 65: 2673–2679.

Bunde, Armin, 1998. Indication of a universal persistence law governing atmospheric variability. *Physical Review Letters*, Vol. 81 No. 3: 729–732.

Casaubon, Meric, 1659. *A True and Faithful Relation of What passed for many Yeers Between Dr John Dee and Some Spirits: Tending (had it succeeded) To a General Alteration of most States and Kingdomes in the World*, 1974 facsimile of the 1659 edition made from a copy in the British Museum. Askin, London.

Coveney, Peter, and Highfield, Roger, 1995. *Frontiers of Complexity*. Ballantine Books, New York.

Day, Michael, 1996. Did ancient Athenians catch Ebola? *New Scientist*, 29 June 1996, Vol. 150 No. 2036: 5.

Deamer, David, 1997. The first living systems: a bioenergetic perspective. *Microbiology and Molecular Biology Reviews*, Vol. 61 No. 2: 239–61.

Diamond, Jared, 1998. *Guns, Germs and Steel*. Vintage, London.

Farrell, Winslow, 1993. *How Hits Happen*. HarperCollins, London.

Freeland, S., and Hurst, L. D., 1998. Load minimization of the genetic code: History does not explain the pattern. *Proceedings of the Royal Society (London)* B, Vol. 265: 2111–2119.

French, Peter, 1972. *John Dee: The World of an Elizabethan Magus*. Ark, London.

Frette, Vidar, Christensen, Kim, Malthe-Sørenssen, Anders, Feder, Jens, Jøssang, Torstein, and Meakin, Paul, 1995. Avalanche dynamics in a pile of rice. *Nature*, Vol. 379: 49.

Gallant, Roy, 1979. *The Constellations: How They Came to Be*. Four Winds Press, New York.

Gardner, Martin, 1978. 'Mathematical games'. *Scientific American*, April 1978: pp. 16–32.

Gleick, James, 1988. *Chaos: Making a New Science*. Heinemann, London.

Goodwin, Brian, 1994. *How the Leopard Changed its Spots*. Weidenfeld and Nicolson, London.

Gopikrishnan, P., Meyer, M., Amaral, L. A. N., and Stanley, H. E., 1998. Inverse cubic law for the distribution of stock price variations. *European Physics Journal* B, Vol. 3: 139–140.

Gouyet, Jean Francois, 1996. *Physics and Fractal Structures*. Masson, Paris.

Gurshtein, Alexander A., 1997. The origins of the constellations. *American Scientist*, Vol. 85, May–June: 264–273

Hart, P., and Oulton, N., 1995. Growth and size of firms. Report prepared for the National Institute of Economic and Social Research.

Hausdorff, J. M., Edelberg, H. K., Mitchell, S. L., Goldberger, A. L., Wei, J. Y., 1997. Increased gait unsteadiness in community-dwelling elderly fallers. *Archives of Physical Medicine and Rehabilitation.*, Vol. 78: 278–83

Held G. A., Solina, D. H., Keane, D. T., Haag, W. J., Horn, P. M., and Grinstein, G., 1990. Experimental study of critical mass fluctuations in an evolving sandpile. *Physical Review Letters*, Vol. 65: 1120.

Holldobler, Bert, and Wilson, Edward O. 1994. *Journey to the Ants*. Belknapp Press, Cambridge, Massachusetts.

Holmes, B., 1995. Message in a genome? *New Scientist*, 12 August, Vol. 147 No. 1990: 30.

Holmes, B., 1998. Irresistible illusions. *New Scientist*, 5 September, Vol. 159 No. 2150: 32–37.

Huberman, B. A., Pirolli, P. L. T., Pitkow, J. E., and Rajan, M. J., 1998. Strong regularities in World Wide Web surfing. *Science*, Vol. 280: 95–97.

Istrail, Sorin, 2000. Universality of intractability for the partition function of the Ising model across non-planar lattices. *Proceeding of the 31st ACM Annual Symposium on the Theory of Computing*: 87–96, 21–23 May 2000, Portland, Oregon. ACM Press, New York.

Jaeger, H. M., Liu, C., and Nagel, S. R., 1989. Relaxation of the angle of repos. *Physical Review Letters*, Vol. 62, 40.

Jaeger, H. M., and Nagel, S. R., 1992. Physics of the granular state. *Science*, Vol. 255: 1523.

Janssen, H. K., Oerding, K., van Wijland F., and Hillhorst, H. J., 1999. Lévy-flight spreading of epidemic processes leading to percolating clusters. *European Physical Journal* B, Vol. 7: 137–149.

Jensen, Henrik Jeldtoft, 1998. *Self-Organised Criticality*. Cambridge University Press, Cambridge.

Jones, Steve, 1993. *The Language of the Genes*. HarperCollins, London.

Jung, P., 1997. Thermal waves and self organization in excitable media. *Physical Review Letters*, Vol. 78: 1723.

Jung, P., Cornell-Bell, A., Madden, K., and Moss, F., 1998. Noise induced spiral waves in *Astrocyte syncytia* show evidence of self-organized criticality. *Journal of Neurophysiology*, Vol. 79: 1098.

Kadanoff, L. P., 1966. Scaling laws for Ising models near T_c. *Physics*, Vol. 2 No. 6: 263–272.

Kadanoff, L. P., 1976. Scaling, universality and operator algebras. In: *Phase Transitions and Critical Phenomena*, Vol. 5A, ed. C. Domb and M. S. Green, pp. 1–34. Academic Press, New York.

Kadanoff, L. P., 1999. *From Order to Chaos II Essays: Critical, Chaotic and Otherwise*. World Scientific, Singapore.

Kadanoff, L., Götze, W., Hamblein, D., Hecht, R., Lewis, E. A. S., Palciauskas, V. V., Martin, R., Swift, J., Aspnes, D., and Kane, J., 1967. Static phenomena near critical points: Theory and experiment. *Reviews of Modern Physics*, Vol. 39 No. 3: 395–431.

Kauffmann, Stuart, 1993. *The Origins of Order*. Oxford University Press, Oxford.

Keitt, T. H., and Stanley, H. E., 1998. Dynamics of North American breeding bird populations. *Nature*, Vol. 393: 257–260.

Kelso, J. A. S., and Hakken, H., 1995. New laws to be expected in the organism: Synergetics of brain and behaviour. In: *What is Life: The Next Fifty Years*, ed. Michael P. Murphy and Luke A. J. O'Neill. Cambridge University Press, Cambridge.

Kephart, Jeffrey O., Hanson, James E., and Siaramesh, Jakka, 1998. Price-war dynamics in a free-market economy of software agents. In: *Artificial Life VI: Proceedings of the Sixth International Conference on Artificial Life*, ed.

Christoph Adami, Richard K. Belew, Hiroai Kitano, and Charles E. Taylor, pp. 53–62. MIT Press, Cambridge, Massachusetts.

Laherrere, J., and Sornette, D., 1998. Stretched exponential distributions in nature and economy: 'Fat tails' with characteristic scales. *European Physical Journal* B, Vol. 2: 525–539.

Lam, L., 1998. *Nonlinear Physics for Beginners*. World Scientific, Singapore.

Langton, Christopher, 1992. Life at the edge of chaos. In: *Artificial Life II, SFI Studies in the Sciences of Complexity*, Vol. X, ed. C. G. Langton, C. Taylor, J. D. Farmer, and S. Rasmussen, pp. 41–91. Addison-Wesley. Redwood City, California.

Levy, Steven, 1992. *Artificial Life: The Quest for a New Creation*. Penguin, London.

Lewin, Roger, 1993. *Complexity*. Dent, London.

Lewin, Roger, 1998. Family feuds. *New Scientist*, Vol. 158 No. 2118: 36–40.

Maddox, John, 1992. Long-range correlations within DNA. *Nature*, Vol. 358: 103.

Mandelbrot, Benoit, 1963. The variation of certain speculative prices. *Journal of Business*, Vol. 36: 394–419.

Mandelbrot, Benoit, 1967. How long is the coast of Britain? Statistical self-similarity and fractional dimension. *Science*, Vol. 155: 636–638.

Mandelbrot, Benoit, 1977. *The Fractal Geometry of Nature*. Freeman, New York.

Mandelbrot, Benoit, 1998. Is nature fractal? *Science*, Vol. 279: 783–784.

Mantegna, R. N., and Stanley, H. E., 2000. *An Introduction to Econophysics*. Cambridge University Press, Cambridge.

Margulis, Lynn, and Sagan, Dorion, 1986. *Microcosmos*. University of California Press, Berkeley, California.

Matthews, Robert, 1999. The power of one. *New Scientist*, 10 July, Vol. 163 No. 2194: 27–30.

Merritt, John I., 1985. *Baronets and Buffalo: The British Sportsmen in the American West, 1833–1881*. Mountain Press, Missoula, Montana.

Miller, George, 1956. The magic number seven plus or minus two: Some limits on our capacity for processing information. *Psychological Review*, Vol. 63 No. 2: 81–97.

Miller, George, 1962. *Psychology: The Science of Mental Life*. Penguin, London.

Munson, P. J., Taylor, R. C., and Michaels, G. S., 1992. Long range DNA correlations extend over entire chromosome. *Nature*, Vol. 360: 636.

Newman, M. E. J., 1997. A model of mass extinction. *Journal of Theoretical Biology*, Vol. 189: 235–252.

Newman, M. E. J., 1999. Small worlds. Unpublished preprint.

O'Brien, K. P., and Weissman M. B., 1994. Statistical characterisation of Barkhausen noise. *Physical Review* E, Vol. 50 No.5: 3446–3452.

Perkovic, O., Dahmen, K., and Sethna, J., 1995. Avalanches, Barkhausen noise, and plain old criticality. *Physical Review Letters*, Vol. 75 No. 24: 4528–4531.

Perrett D..I., May K., and Yoshikawa, S., 1994. Attractive characteristics of female faces: Preference for non-average shape. *Nature*, Vol. 368: 239–242.

Petzinger, T. J., 1997. *How light bulbs, Hootie contributed to predicting change*. *Wall Street Journal*, 28 March.

Ray, T. S., 1976. Skototropism and the natural history of some tropical vines. Honors thesis in biology, Florida State University.

Sartre, Jean-Paul, 1948. *Existentialism and Humanism*. Methuen, London.

Savaglio, Sandra, and Carbone, Vincenzo, 2000. Scaling in athletic world records. *Nature*, Vol. 404: 244.

Schama, Simon, 1996. *Landscape and Memory*. Fontana, London.

Schroeder, Manfred, 1991. *Fractals, Chaos, Power Laws: Minutes from an Infinite Paradise*. Freeman, New York.

Shapir, Y., Raychaudhuri, S., Foster, D. G., and Jorne, J., 2000. Scaling behaviour of cyclical surface growth. *Physical Review Letters*, Vol. 84 No. 14: 3029–3032.

Shumaker, Wayne, 1982. *Renaissance Curiosa*. Medieval and Renaissance Texts, New York.

Smith, Matthew, Wiseman, Richard, Harris, Peter, and Joiner Richard, 1996. On being lucky. *European Journal of Parapsychology*, Vol. 12: 35–43.

Solé, R., Manrubia, S. C., Benton, M., Kauffmann, S., and Bak, P., 1999.

Criticality and scaling in evolutionary ecology. *Trends in Ecology and Evolution*, Vol. 14 No. 4: 156–160.

Somfai, E., Czirok, A., and Vicsek, T., 1994. Self-affine roughening in a model experiment on erosion in geomorphology. *Journal of Physics* A, Vol. 205: 355.

Spinney, L., 1998. I had a hunch . . . *New Scientist*, 5 September, Vol. 159 No. 2150: 42–47.

Stanley, H. E., 1969. Critical indices for a system of spins of arbitrary dimensionality situated on a lattice of arbitrary dimensionality. *Journal of Applied Physics*, Vol. 40 No. 3: 1272–1274.

Stanley, H. E., Afanasyev, V., Amaral, L. A. N., Buldyrev, S. V., Goldberger, A. L., Havlin, S., Leschorn, H., Maass, P., Mantegna, R. N., Peng, C.-K., Prince, P. A., Salinger, M. A., Stanley, M. H. R., Viswanathan, G. M. 1996a. Anomalous fluctuations in the dynamics of complex systems: From DNA and physiology to econophysics. *Physica* A, Vol. 224: 302–321.

Stanley, H. E., Amaral, L. A. N., Buldyrev, S. V., Goldberger, A. L., Havlin, S., Leschorn, H., Maass, P., Makse, H. A., Peng, C.-K., Salinger, M. A., Stanley, M. H. R., Viswanathan, G. M. 1996b. Scaling and universality in animate and inanimate systems. *Physica* A, Vol. 231: 20–48.

Stanley, H. E., Buldyrev, S. V., Goldberger, A. L., Havlin, S., Mantegna, R. N., Chung-Kang, P., and Simons, M., 1996. Scale invariant features of coding and noncoding DNA sequences. In: *Fractal Geometry in Biological Systems: An Analytical Approach*, ed. P. M. Iannaccone and M. K. Khoka, pp. 15–30. CRC Press, Boca Raton, Florida.

Stanley, H. E., Coniglio, A., Klein, W., Nakanishi, H., Redner, S., Reynolds, P. J., and Shlifer, G., 1980. Critical phenomena: Past, present and 'Future'. In: *Dynamics of Synergetic Systems: Proceedings of the International Symposium on Synergetics*, ed. H. Haken, pp. 22–38. Springer. Berlin.

Stewart, Ian, 1990. *Does God Play Dice?* Penguin, London.

Strogatz, Steven H., and Stewart, Ian, 1993. Coupled oscillators and biological synchronization. *Scientific American*, December, Vol 269 No. 6, pp. 68–75.

Strong, D. R., and Ray, T. S., 1975. Host tree location behavior of a tropical vine (*Monstera gigantea*) by skototropism. *Science*, Vol. 190: 804–806.

Suki, B., Alencar, A. M., Sujeer, M. K., Lutchen, K. R., Collins, J. J., Andrade, J. S., Ingenito, E. P., Zapperi, S., Stanley, H. E., 1998. Life support system benefits from noise. *Nature*, Vol. 393: 127–128.

Taylor, R. P., Micolich, A. P., and Jonas, D., 1999. Fractal analysis of Pollock's drip paintings. *Nature*, Vol. 399: 422.

Thomas, Keith, 1971. *Religion and the Decline of Magic*. Penguin, London.

Thomas, Keith, 1984. *Man and the Natural World*. Penguin, London.

Voss, R., 1992. Evolution of long-range fractal correlations and $1/f$ noise in DNA base sequences. *Physical Review Letters*, Vol. 68: 3805–3808.

Voss, R., and Clark, J., 1978. $1/f$ noise in music: Music from $1/f$ noise. *Journal of the Acoustical Society of America*, Vol. 63: 258–263.

Waldrop, M. Mitchell, 1992. *Complexity*. Penguin, London.

Watson, Peter, 1981. *Twins: An Investigation into the Strange Coincidences in the Lives of Separated Twins*. Hutchinson, London.

Weibel, Ewald R., 1993. Design of biological organisms and fractal geometry. In: *Fractals in Biology and Medicine*, ed. T. F. Nonnenmacher, G. A. Losa and E. R. Weibel, pp. 68–85. Birkhauser, Basle.

Wertheim, Margaret, 1997. *Pythagoras' Trousers: God, Physics and the Gender Wars*. Fourth Estate, London.

West, G. B., Brown, J. H. and Enquist, B. J., 1997. A general model for the origin of allometric scaling laws in biology. *Science*, Vol. 276: 122–126.

Wilson, Kenneth G., 1983. The renormalization group and critical phenomena. *Reviews of Modern Physics*, Vol. 55 No. 3: 583–600.

Wolfram, Stephen, 1984. Cellular automata as models of complexity. *Nature*, Vol. 311 No. 4: 419–424.

Wright, Lawrence, 1997. *Twins*. Weidenfeld and Nicolson, London.

Yam, Philip, 1995. Talking trash. *Scientific American*, March, Vol. 272 No. 3: 12.

Yates, Frances A., 1964. *Giordano Bruno and the Hermetic Tradition*. University of Chicago Press, Chicago.

Index

Acanthamoeba castellanii, 131
Adami, Christoph, 106, 108–11
Advanced Telecommunications
    Research laboratories,
    Kyoto, 106
After Many a Summer Dies the Swan
    (Huxley), 219
ageing, 146
Agrippa, Henry Cornelius, 11
albatross, 165–7
Albuquerquia, 193
alpha waves, 251
Amaral, Luis, 216–18
American Museum of Natural
    History, 174
amino acids, 126–7
Amy (twin), 265–6
anatomical structures, 156–60
*Anatomical Treatise on the Motion of
    the Heart and Blood in
    Animals, An* (Harvey), 142
Anderson, Peter, 43
Andrews, Thomas, 57
angels, summoning, 10–13
animals
    extinction, 173–6
    foraging, 166–7
    group behaviour, 168–71
    markings, 158
    movements, 165–8
    organ structures, 158
    population size, 172–3

antibiotics, 34
anticipation, 246
ants
    colonies, 168–70
    foraging, 167
applause, 239–40
Aristotle, 41, 141
art, 155–6, 243–4
'Art of Fugue' (Bach), 150
Arthur, Brian, 193
artificial life, 100–1, 104–5
Ash Wednesday Supper, The (Bruno),
    9
astrobiology, 95–6
astrology, 22–3, 26–30
astronomy, 18, 26–30
At Home in the Universe (Kauffman),
    88
athletic performance, 241
Auden, W. H., 1
'Auguries of Innocence' (Blake), 43
autocatalytic sets, 122
avalanches, 53–4, 55, 85, 86, 87, 113,
    115–17
Avida, 106, 108–10
Avnir, David, 161–2

Bach, Johann Sebastian, 150–1
Bacon, Kevin, 229–30
bacteria, 93, 122–3, 136–7
    growth, 34
Bak, Per, 86, 111, 113, 115, 117, 175,
    278

INDEX

Barabási, Albert-Lászió, 239–40
Barnsley, Michael, 278
Barrett, Chris, 193
bases, 126
Bass, Thomas, 205
Bath University, 133
beauty, 154–6
Beethoven, Ludwig van, 151–2
beliefs, 33
bell curve, 80
Benford's Law, 213–15
Bergson, Henri, 245
beta waves, 251–2
Beth (twin), 265–6
Beth Israel Deaconess Medical Center, Boston, 147
Beysens, Daniel, 118
Bible, 17
Biham, Ofer, 161–2
biology, 291–2
birds
    albatross movements, 165–7
    population size, 172
Birkhoff, George David, 154
Black Death, 222, 226
Blake, William, 43
blood, 89
blood vessels, 156–7, 159–60
blue-green algae, 93, 122
body organs, 156–60
Boltzmann, Ludwig, 62, 63
Bosch, Hieronymous, 4–5
Boston Museum of Science, 148
Boston University, 129, 166, 201
Bouchard, Tom, 263
Boyle, Robert, 23
Boyle's law, 57
brain, 249–53, 272–3
    cell types, 252
    imaging, 253
    information processing, 258–60
    waves, 146, 251–3, 259
breast cancer, 269
breathing, 145
Bristol University, 167
bronchi, 157, 159–60
Brookhaven National Laboratory, 111
Brown, Titus, 106
Browne, James, 159

Bruno, Giordano, 9, 23
Bubonic plague, 222, 226
buffalo, 170–1
Bunde, Armin, 292–3
Burgess, Anthony, 274
butter, colour of, 33
buyer behaviour, 177–89

Caenorhabditis elegans, 132
California Institute of Technology, 106
California University, 148
Cambrian explosion, 175
cancer, 269
carbon dioxide, 57
Carbonet, Vincenzo, 241
Castelldefells, Catalonia, 43–50
Catalonia University, 124, 170
catastrophe theory, 37–8
cellular automata, 98–103, 106–11, 170
centromeres, 128
cerebellum, 252
chaos, 101–2, 105, 119, 254, 275–6
chemical complex networks, 121
Chicago University, 65, 115, 175
chunking, 259
circulation, 156–7, 159–60
city populations, 79–80
clapping, 239–40
Clare, 44, 45
Clarke, John, 148
Clarke, Samuel, 281–2
clouds, 256–7
coastline length, 81–2
codon, 126
colonies, 168–70
coma, 253
Comedians, The (Greene), 9
company growth, 215–18
complexity, 114, 118, 276
computer analogy, 283–4
condensation, 118
condensed matter, 111
constellations, 26–8
Conway, John Horton, 98–9
Copernicus, Nicolaus, 18, 23
Cornell University, 69, 72, 287
correlation length, 74

INDEX

Cortés, Hernando, 220, 226
cotton prices, 196–7
Cracked Rear View, 185
Craig, John, 281
criminal behaviour, 242–3
critical exponent, 78
critical point, 54, 57–60, 63, 76, 78
crowd dynamics, 235–7
Cuitláhuac, emperor, 226
Curie point, 59
cytomegalovirus, 125

Darwin, Charles, 174, 282
Davids, Zach, 147
Dee, John, 10–13, 14–16, 26
delta waves, 251
Department of the Environment,
    Transport and the Regions,
    237–8
Die Kunst der Fugue (Bach), 150
dimensions, 73–4
dinosaurs, 173, 176
Diomedea exulans, 165–7
*Discourse Concerning the
    Unchangeable Obligations
    of Natural Religion, A*
    (Clarke), 281–2
disease
    inheritance, 269
    spread, 219–28
Ditto, Bill, 144
DNA, 123, 125–35
Domb, Cyril, 65
doolally, 232–3
Drosophila melanogaster, 131
Dudley, Robert, 14
Dyson, Freeman, 72

Earthly Powers (Burgess), 274
earthquakes, 55
    prediction, 33
Ebola virus, 222, 225
Eciton burchelli, 167
École Centrale, 117
economics, 209–12
econophysics, 195–208
'edge of chaos', 101, 102, 105
Egypt, ancient, 28
Einstein, Albert, 31

El Farol problem, 193–5
Eldridge, Niles, 174
Eliot, T. S., 274
Elizabeth I, 15
Emergent Solutions Group, 183–7
Enquist, Brian, 159
epidemics, 219–28
epilepsy, 252–3
error catastrophe, 124
Eudaemonic Pie, The (Bass), 205
Europa, 95–6
Evesque, Pierre, 117
existentialism, 42
Existentialism and Humanism (Sartre),
    42
exons, 127
exponential growth, 288
extinction, 173–6
*Extraordinary Popular Delusions and
    the Madness of Crowds*
    (Mackay), 177
extraterrestrial life, 95–6

facial attractiveness, 154–6
Fang, Forrest, 153–4
Farmer, Doyne, 204–5
Farrell, Winn, 183–7, 189
Fass, Craig, 230
fat-tailed distribution, 223–4
fate, 260
Feder, Jens, 116
Feigenbaum, Mitchell, 254, 256–7
Ferdinand, Archduke Franz, 234
Feynman, Richard, 72
Ficino, Marsilio, 20, 21
financial markets, 55, 195–204,
    206–8, 211
Finnegan, Judy, 267
fireflies, 167–8
fires, escape from, 236
fish schools, 171
Fisher, Michael, 65, 69
Florida Atlantic University, 253
foraging, 166–7
fossil records, 172–4
Fractal Geometry of Nature, The
    (Mandelbrot), 84
fractals, 78–85
    anatomical structures, 156–60

INDEX

faces, 155–6
music, 149–54
Sierpinski carpet, 149–50
Franks, Nigel, 167
fraud, 212–15
Freeland, Steven, 133
Freud, Sigmund, 283
fruit fly, 131
'fuck-up factor', 217–18
Full Monty, The, 183

Gaia, 284
gait, 164–5
gambling, 205
Game of Life, 99
Garden of Earthly Delights (Bosch), 5
Gardner, Martin, 149
gases, 57
GenBank, 130
gene
    inheritance, 265–73
    networks, 120–1
General Electric Company, 213
Georgia Institute of Technology, 144, 252
Gibrat, R., 215
Giessen University, 292
Ginelli, Mike, 230
Gleick, James, 276
Goldberger, Ary, 144, 147
Goodwin, Brian, 170
Gore, Al, 278
Gould, Stephen Jay, 174
'Great Chain of Being', 19
Great Plague of Athens, 222
Greece, ancient, 30
Green, Mel, 65, 70
Greene, Graham, 9
group organization, 168–71
Guare, John, 219
Gulf Stream, 94
gut micro-organisms, 280

Harvard University, 174, 229
Harvey, William, 23, 141–2
heart, fractal structure, 156–7
heartbeat, 55, 138–45
    artificial pacemakers, 144
    defibrillators, 144–5

fibrillation, 144
heart disease, 143–4
interbeat interval, 142–4
music, 147–8
Heartsongs, 147
Hebb, Donald, 247
Hegel, Georg W. F., 234
height, 269–70
Held, Glen, 116
Henry VIII, 17
Hermes Trismegistus, 21, 23
Hermeticism, 21–6
Higashi, Seigo, 169
hippocampus, 252
history, 233–4
History of the Peloponnesian War (Thucydides), 222
hit creation, 183–9
Hokkaido University, 169
Hootie and the Blowfish, 185–7
Houthakker, Hendrik, 196
Huberman, Bernardo, 238
Human Behaviour and the Principle of Least Effort (Zipf), 79
humanism, 24
hunches, 277
Huntington's disease, 269
Hurst, Lawrence, 133
Huxley, Aldous, 219
Huyssen, Joachim, 166
hydrobot, 96
hydrogen bonds, 89–90

IBM research centres, 116, 148
ice, 96–7
IDSIA, 155
Illinois University, 287
influenza, 227–8
information spread, 186, 230–1
information theory, 103
inheritance, 265–73
insect spiracles, 158
interference, 103–4
Internet, 238–9
introns, 127, 129, 132
intuition, 245–6
Ishikari Bay, 169
Ising model, 65, 289–90
Istrail, Sorin, 290

Jaeger, Heinz, 115
Jensen, Henrik Jeldtoft, 118
Jesus Christ
    Christian symbol for, 29
    Second Coming, 281
Jim twins, 261–3
Jorné, Jacob, 157–8
Jøssang, Torstein, 116
Jung, Peter, 252
Jupiter's satellites, 95–6

Kadanoff, Leo, 65–9, 70, 115, 286
Kauffman, Stuart, 88, 119–22, 135
Keitt, Timothy, 172
Kelley, Edward, 10–13
Kelso, Scott, 253, 258
Kennerley, William Clark, 170–1
Keynes, John Maynard, 177
Klein, Gary, 245–6, 277
Knobler, Charles M., 118

Landau, Lev Davidovitch, 63, 64
Landscape and Memory (Schama),
    231
Langton, Chris, 100–3, 104–5, 170
language, 80, 232–3
Laplace, Pierre Simon de, 34
Larkin, Philip, 88
law of large numbers, 210
learning, 33
LeDoux, Joseph, 246
Legion, 237
Lenski, Richard, 106
Leonardo da Vinci, 155–6
Levy distribution, 201
Lewinsky, Monica, 230
Lewis, James, 261–3
Lidar, Daniel, 161–2
life, origin of, 88–137
lipids, 91
Little, David Clark, 153
Little Gidding (Eliot), 274
Longitude (Sobel), 183
Los Alamos National Laboratory,
    159, 193, 257
Louvain, 11
Lovejoy, Sean, 256
Lovelock, James, 284
loyalty cards, 180–1

lungs, 157, 159–60
Luther, Martin, 17

McGill University, 247
Mackay, Charles, 177
Madeley, Richard, 267
Madimi, 12
magic, 25
magnets, 59–60, 63–9, 83
magus, 10–11
Malcai, Ofer, 161–2
Mallon, Mary, 224–5
Mandelbrot, Benoit, 78, 79, 81, 82–4,
    162, 196–7, 284–5
Mantegna, Rosario, 200
Mars, 95
Marx, Karl, 282–3
Massachusetts Institute of Technology
    (MIT), 167, 277
Mathematical Principles of Christian
    Theology (Craig), 281
Maxwell, James, 62
May, Robert, 254, 255–6
mean field theory, 61–7
Medici, Cosimo di, 20
memes, 230–4
memory, 259
messenger-RNA, 132–3
metaphysics, 40–1
Metaphysics (Aristotle), 41
Metaphysics (Taylor), 1
mice, spontaneous creation, 33
Michelangelo, 29
Michigan State University, 106
Migdal, Sasha, 70
Milgram, Stanley, 229
Miller, George, 258–9
Minnesota University, 263
Miramontes, Octavio, 170
Mona Lisa, 156
monkeys, typewriting, 285
More is Different (Anderson), 43
Mortlake, 9
Moses (sculpture), 29
motor neurons, 252
movement, 164–8
mud piles, 117
Murifri, 12
music, 147–54

record sales, 185–8, 290
myocytes, 157
myosin heavy chain protein, 131–2

Nagel, Sidney, 115
name popularity, 267
National Bureau of Standards, 70
nature/nurture debate, 265–73
naval terms, 232–3
nervous system
    conduction, 250–1
    connections, 252
    parasympathetic/sympathetic,
      140–1
    see also brain
Neumann, John von, 98
New Mexico University, 159
Newcomb, Simon, 213
Newton, Isaac, 23, 33, 34, 280–1, 282
Nice University, 288
Nigrini, Mark, 214
noise, 103–4
    $1/f$, 84–5, 111–12, 149, 287
normal distribution, 80
Notre Dame University, 239
Novalis, 138
nucleotides, 125–6

O'Brien, Kevin, 287
oceans, 92–4
    currents, 94
'Ode to Terminus' (Auden), 1
Ofria, Charles, 106
On Growth and Form (Thompson),
    138
On the Revolution of the Heavenly
    Spheres (Copernicus), 18,
    23
Onsager, Lars, 65, 289
Open University, 170
optical illusions, 245
Orchid Fractal Engineering, 236
order, 36–7
order parameter, 74
organ structures, 156–60
Organised Chaos (Thompson), 153
Orion, 242–3
Oslo University, 116

Packard, Norman, 204–5
Palo Alto Research Center, 238
parasympathetic nervous system,
    140–1
Pareto, Vifredo, 196
particle physics, 4
Patashinskii, Aleksandr Z., 69
pattern spotting, 32–5
Penicillium notatum, 34
Pennsylvania University, 120
perception, 246–8
percolation, 92–3
Perrett, David, 154
personality traits, 270–1
phase transitions, 58–9, 64, 65–6, 72,
    253
physical attraction, 154–6
'Physical Observation No. 248', 138
physics, 3–4, 5–6, 30, 31, 36, 291
Pimander, 21
Pippard, Brian, 65
Pizzaro, Francisco, 220
plague of Antoninus, 222
plague of Justinian, 222
planets, 18, 28–9, 30, 95–6
Plough constellation, 27
Pokrovskii, Valerii L., 69
Pollack, Jackson, 243–4
Polyakov, Sasha, 70
population size, 79–80, 172–3, 254–6
power law, 85, 86, 117, 252
precession, 29
predictability, 33
Prediction Company, 205–6
Predictors, The (Bass), 205
Pretoria University, 166
PricewaterhouseCoopers, 182–7
printing, 17
Proffitt, Dennis R., 246
prokaryotes, 93, 122–3, 136–7
proteins, 123, 127, 131–2, 136
psychoanalysis, 283
punctuated equilibrium, 174–5
purines, 126
Purkinje cells, 252
pyrimidines, 126
Pythagoras of Samos, 30

rainfall, 257
random walk, 129–30
Raup, David, 174–5
Ray, Tom, 106–8
reaper program, 108
record sales, 185–8, 290
reductionism, 35
Reformation, 16–17
regularities, 5
religion, 16–17, 21, 31, 40, 281–2
Renaissance, 18–19, 20, 23, 25, 26
renormalization, 72, 86
respirators, 145
rhythms, 38
ribosomes, 123
rice piles, 116
Richardson, Lewis, 284
RNA, 122–4, 125
    messenger-RNA, 132–3
Robbins, Lionel, 209
robot submarine, 96
Rochester University, 157
Rossmo, Kim, 242
roulette, 205
Russell, Bertrand, 245

Saccharomyces cerevisiae, 131
Sainsbury's, 177–82
St Andrews University, 154
sand
    constructions, 51–2
    piles, 52–4, 86, 87, 113, 115–16,
        117, 286–7
Sandia National Laboratory, 290
Santa Fe Institute, 100, 120, 172, 193
Sartre, Jean Paul, 42
Savaglio, Sandra, 241
scale invariance, 213–14
Schama, Simon, 231
Schillinger, Joseph, 148, 284
Schmidhuber, Jürgen, 155
Schwinger, Julian, 72
science, 24–32, 33–4, 275–6, 293
seances, 10–13
Second Coming, 281
self-organized criticality, 86–7, 112–13,
        115–18, 175, 278, 279, 287
self-reflection, 32
self-similarity, 82–3

sensory deprivation, 247–8
Sepkowski, John, 175
Sethna, James, 287
Shannon, Claude, 103, 104
Shapir, Yonathan, 157–8
shell patterns, 158–9
Shibam, 52
shopping behaviour, 177–82
Sidney, Philip, 14
Sierpinski carpet, 149–50
SimStore, 177–82
sinoatrial node, 140
six degrees of separation, 229–30
Six Degrees of Separation (Guare),
        219
sleep apnoea, 143
sleeping sickness, 222
smallpox, 222, 227
snail shells, 158
Sobel, Dava, 183
social groups, 168–71
social networks, 185, 228–30
Solé, Ricard, 123–4, 170
Sony Computer Science Laboratories,
        208
Soper, George, 224
Sornette, Didier, 288
Southern Methodist University, Dallas,
        214
Spengler, Oswald, 233–4
Spinoza, Baruch, 245
sport, 241
Springer, James, 261–3
SQUID, 253
Standard and Poor's 500 index, 200,
        290–1
Stanley, Gene, 129–32, 163, 195,
        200–2
Staphylococcus aureus, 34
stars, 26–8
statistical mechanics, 61–2, 63
statistical theories, 114
steady state, 54
Stewart, Ian, 153
Still, Keith, 235–6
stock market, 55, 195–204, 206–8,
        211
strange attractors, 257
stretched exponential, 288

Strogatz, Steven, 167–8
stromatolites, 137
supercolonies, 169–70
supermarket models, 177–82
Swiss Bank Corporation, 206
sympathetic nervous system, 140–1
systems, 54, 55–7

Takayasu, Hideki, 208
Tang, Chao, 86, 111
Taylor, Richard, 1
telomeres, 128
theology, 281
Theory of Everything, 3
thermodynamics, 59–60, 62, 65
Thom, René, 37–8
Thompson, D'Arcy, 138
Thompson, Phil, 153
Thucydides, 222
Tierra, 107–8
town models, 193
Toynbee, Arnold, 233
Tract on Monetary Reform, A
      (Keynes), 177
Trades and Quotes Database, 201
traffic jams, 237–8
train travel, 297
trypanosomiasis, 222
tumour growth, 158
Turtle, Brian, 230
twin studies, 261–8
Tyndale, W., 17
typewriting monkeys, 285
Typhoid Mary, 224–5

Ulam, Stanislaw, 98
universality, 3, 4, 70–1, 77–8, 276–9,
      293–8
universe, 18

Van der Waals, 57
Venables, Mark, 181
Vicsek, Tomas, 117

Virginia Commonwealth University,
      263–4
Virginia University, 246
viruses, 121–4
visual illusions, 245
Viswanathan, Gandimohan, 166–7
Voronel, Sasha, 65
Voss, Richard, 125, 148
Vostok, Lake, 96

walking, 164–5
Ward, Mark, observations on beach,
      44–50
water, 88–9, 95–7
  boiling point, 57–8
  chemical structure, 88–90
  heat capacity, 94
  ice, 96–7
'Water' (Larkin), 88
Watson Research Center, 148
waves, 46, 49
weather patterns, 292–3
Weiss, Pierre, 63
Wembley Stadium, 235, 236
West, Geoffrey, 159
West Nile Virus fever, 227
white blood cells, 146
Widom, Ben, 69
Wiesenfeld, Kurt, 86, 111
Wilson, E. O., 106
Wilson, Kenneth, 72, 73–5, 77, 286
Wolfram, Stephen, 100
word frequency, 80
world soul, 22
World Wide Web, 238–9

Xerox Palo Alto Research Center, 238

Yates, Frances, 20
Yorktown Heights, 116

Zipf's law, 79–80, 285
zombie, internal, 244–5